Rethinking Peace and Conflict Studies

Series Editor
Oliver P. Richmond
University of Manchester
Manchester, UK

W0111819

This agenda-setting series of research monographs, now more than a decade old, provides an interdisciplinary forum aimed at advancing innovative new agendas for approaches to, and understandings of, peace and conflict studies and International Relations. Many of the critical volumes the series has so far hosted have contributed to new avenues of analysis directly or indirectly related to the search for positive, emancipatory, and hybrid forms of peace. New perspectives on peacemaking in practice and in theory, their implications for the international peace architecture, and different conflict-affected regions around the world, remain crucial. This series' contributions offers both theoretical and empirical insights into many of the world's most intractable conflicts and any subsequent attempts to build a new and more sustainable peace, responsive to the needs and norms of those who are its subjects.

More information about this series at
http://www.palgrave.com/gp/series/14500

Susan Forde

Movement as Conflict Transformation

Rescripting Mostar, Bosnia-Herzegovina

Susan Forde
Department of Politics
University of York
York, UK

Rethinking Peace and Conflict Studies
ISBN 978-3-030-06483-9   ISBN 978-3-319-92660-5   (eBook)
https://doi.org/10.1007/978-3-319-92660-5

Cover credit: Cover art by Mišela Boras

Printed on acid-free paper

This Palgrave Macmillan imprint is published by the registered company Springer International Publishing AG part of Springer Nature
The registered company address is: Gewerbestrasse 11, 6330 Cham, Switzerland

For my Gran, Ann Littler, a wonderful woman.

Preface

Constellations of Transformation

Space, memory, and time create constellations of movement. We are born into space and time, and become social beings through memory and movement, experiences of space at different intersections of time produces memories. Similarly, as space changes over time, memories change over time. Others we meet in space are integral to our memories and understanding, not only of ourselves, but of the space itself. Our memories and experiences of space effects perceptions of, and therefore expectations of space. Positive experiences in space (what becomes place through communicating the story socially) impact on how we, and others we interact with, think about, and use said space in the future. Negative experiences of space such as traumatic or upsetting events can limit or change movement. This may be during a violent intra-state conflict or in "everyday" life, of course the risk of such negative experiences greatly increases in an intra-state conflict. However, as space can transition from safe to un-safe, so too can it be transformed from un-safe to safe, and spaces can of course invoke seemingly little feeling at all. As there is a multiplicity of factors which effect movement and perceptions of space there can be no one formula for transforming experiences of space. Our experiences of space are highly individual due to our spatial trajectories (or positionalities) as where we are "from", impacts on how we process, understand, and move through space. The picture on the front cover of this book, 'Pick your glasses' is a piece of street art in Mostar by Mišela

Boras. I have used it in presentations about my work in Mostar to explain theory as a way of looking at a situation. This book should very much be seen in the same way, it is a way of looking at Mostar, informed by my own spatial trajectory. It proposes looking at and understanding post-conflict space through spatial narratives of residents. Focusing in on Mostar, Bosnia Herzegovina this book demonstrates the transformative potential of movement and use of space in the city.

The city of Mostar, has at times been overgeneralised in academia, therefore, this work relays a narrative of the city, as one that is still deeply divided, but is transforming every day, in unseen and subtle ways. Though the work centres on Mostar, it presents an analytical framework that can be applied to any spatiality. While the narratives are of movement in Mostar, you may find common experiences within the narratives, though you may never have set foot in the city. Certainly, we have all felt out of place in one context or another, and more so, our experience of space is not static. We know the power of changing experiences of space, of revisiting a favourite place, of returning home after a period of time away, of avoiding a darkened street or park at night, but using the space happily in daylight. So, though this book is grounded in conflict transformation literature, it contributes a framework of analysis for identifying the transformation of space, by drawing on other disciplines, such as human geography and sociology.

This book explores historical, academic, theoretical and social narratives to set the context of the city of Mostar. The forthcoming chapters discuss the spatial dimensions of post-conflict space and reflect on the social use of the city space of Mostar. Through spatialising social narratives of the use of the city space, the work contributes to existing literature on the city of Mostar. The book begins by discussing the main themes of the work, in Chapter 1 you will be introduced to the framework of the book and also the city. The chapter also discusses the application of mapping as a methodology which was complemented by narrative interviews. Chapter 2, spatialises the structure and agency debate and through the theoretical framework demonstrates how social narratives rescript the use of the institutionally restaged city space. The interactivity of the two concepts, (re-)scripting and (re-)staging is also set out. The following Chapter 3 then locates the spatialisation of peace in the four interacting spaces of the institutional, urban, social, and personal. In the institutional space, the actors involved, in the restaging and maintenance of physical locations in the urban environment are discussed. Fundamentally, the dominance of institutional actors in the production of space demonstrates the capability and responsibility of institutional actors to alter the cityscape and stage peace. In the urban context, restaging is performed by institutional actors, in order to direct the social use of space in cities. While cities are typically staged top-down the construction, urban spaces are inherently social spaces which can be transformed by social actors. In Mostar, as social actors have established shared spaces for the purpose of traversing ethno-nationalistic institutional divides, the conflict transformation potential of social actors to rescript the urban environment is proposed. Fundamentally, the capabilities of social actors to influence the use of, and spatial meaning of places in the cityscape are discussed. The social and personal spaces are collapsed in to each other and social actors are presented as agentive, and capable of socially rescripting divisively staged city space for a purpose different from its staging.

Chapter 4, provides a spatialised account of the historical formation of Mostar, charting narratives from initial settlement to recent history of the city. The theoretical conceptualisations of restaging and rescripting are employed to present the institutional establishment, and social use of the city space historically. This chapter reflects on the transgenerational narratives of historical restaging and rescripting of the space, which were instrumental in the contested ethno-nationalistic narratives and directed the 1992–1995 Bosnian War. Additionally, the chapter provides a social narrative of the spatial context of the city through presenting narratives of movement during the 1992–1995 war.

The social narratives of movement provided by participants, is presented across two chapters. Both Chapters 5 and 6 discuss spaces in the city. Chapter 5, discusses the spatial-institutional divisions in Mostar and how participants negotiate the divide. In focus, Chapter 5 discusses how social actors can rescript space (which may clash or correlate with the institutional staging of space) through the visibility of symbols, graffiti, and street art which reflect narratives of peace and conflict. Through contemporary narratives of movement; temporal, functional movement, and socially weighted movement are discussed as influencing social rescripting in the city. These variables of movement demonstrate how convenience, time, and context can direct movement which may be divisive or transformative, in the context of the divided city. This chapter also discusses the divisions in education in Bosnia-Herzegovina (BiH) and Mostar and explores how use of shared spaces has transformed divisional movement.

In Chapter 6, the social narratives regarding the use of spaces of movement in the city are presented. This part of the work travels through popular spaces of movement in the city; and through the narrative of the spaces, the restaging and the social rescripting of the spaces is presented. Through the participant narratives, the work demonstrates the capability of social actors to rescript spaces in the city, though this is typically time and context dependent.

Chapter 7, discusses the implications of the research in Mostar and presents the relevance of spatial agency in the city, and the research, to other divided cities. This chapter discusses the theoretical contribution of the conceptualisations of rescripting and restaging as allowing for a deeper understanding of social dynamics of movement, particularly in divided cities. Furthermore, the chapter illustrates the potential for social movement to rescript divisively staged city space, which also rescripts social relations within space, therefore, enacting conflict transformation. Consequently, the chapter presents an outlook for research in divided cities that is informed by spatial analysis of movement.

This book provides a qualitative analysis of post-peace agreement space. The Dayton Agreement that ended the war in Bosnia-Herzegovina has enshrined conflict divisions, yet through movement social actors demonstrate their agentive capabilities to enact conflict transformation. Fundamentally, as socio-spatial beings our memory is spatialised and is informed by movement through spaces and places. In intra-state conflict, everyday spaces become staged (through destruction or damage) and scripted (through personal experiences in the space) with conflict memories. Therefore, following a conflict, it is important for the city to be restaged through post-conflict reconstruction, which can allow residents of the city to rescript space. Of course, the opinions and experiences of residents of any city are not homogenous and there will exist a wide variety of local voices and opinions concerning post-conflict reconstruction. Accordingly, this book explores social spaces in the city set out by participants, contextualising the spaces through their spatial narratives. Through focusing on the conflict transformation potential of social rescripting in post-conflict environments, this work presents the agentive capabilities of social actors to enact conflict transformation, though this may be subtle, or temporal in nature.

York, UK   Susan Forde

ACKNOWLEDGEMENTS

This work is dedicated to the participants who shared their narratives of Mostar with me. I am deeply grateful to all those who made this research possible by participation. The book originates from my doctoral research which I completed at Liverpool Hope University. I owe sincere thanks to Dr. Stefanie Kappler who provided valuable comments on my work and constant support during this time. I am also grateful for support I received to conduct the research, and for support from colleagues in the Department of History and Politics at Liverpool Hope University. I would especially like to thank Professor Nicholas Rees for his support, and would also like to thank Professor Annika Björkdahl, for comments on earlier drafts of this work. Special thanks are also extended to new colleagues and friends from the Department of Politics at the University of York for their support this past year.

Warm thanks are always reserved for my family and my friends who support me in all I do. Over the course of writing this book I have had a lot of opportunities to script and rescript my own experiences of space. I would like to thank those that have been a part of this journey and for all the conversations we have had exploring space and time together.

CONTENTS

CONTENTS

ABBREVIATIONS

ABiH	Armija Republike Bosne i Hercegovine (Army of the Republic of Bosnia and Herzegovina)
BiH	Bosna i Hercegovina (Bosnia and Herzegovina)
EU	European Union
EUAM	European Union Administration of Mostar
EUFOR	European Union Peacekeeping Force
FBIH	Federation of Bosnia and Herzegovina
HVO	Hrvatsko vijeće obrane (Croatian Defence Council)
IRA	Irish Republican Army
JNA	Jugoslovenska narodna armija (Yugoslav People's Army)
MOU	Memorandum of Understanding
NATO	North Atlantic Treaty Organisation
OHR	Office of the High Representative
OSCE	Organisation for Security and Co-operation in Europe
PMC	Pavarotti Music Centre
RS	Republika Srpska
SAF	Street Arts Festival (Mostar)
SFRY	Socialist Federal Republic of Yugoslavia
UN	United Nations
UNESCO	United Nations Educational, Scientific and Cultural Organisation
UNPROFOR	United Nations Protection Force
UWC	United World College
VRS	Vojska Republike Srpske (Army of the Serb Republic)

LIST OF FIGURES

CHAPTER 1

Space, Time and Mostar: *Welcome to Mostar*

During my second visit to Mostar in April 2015 I was told that the city is so divided taxi drivers will not cross the bridges to the other side of the city. In September 2015, I was with a friend who lived in Mostar we had sat outside where she worked catching up, it was late when we finished talking, and she said I should get a taxi back to my hotel, I mentioned what I had been told previously and we both wondered if the driver would not want to cross the river. When the taxi arrived, the driver told us both 'I will take you to Vienna if you pay me'. While this is somewhat anecdotal, what I aim to demonstrate through this narrative is the existence of generalisations which become social myths. The narrative of divisions so entrenched that they cause taxi drivers (presumably from both sides) to veto fares which cross the river, emphasises the enduring narrative of the intractability of the conflict. This may be the case for some, it may align with a lived experience or one heard in passing, it may once have been a common problem, and indeed for some it may still be. However, it is important to query beyond generalisations regarding the persistence of ethno-nationalistic divisions in the city which have the potential to exacerbate and maintain divisions. Such generalisations do of course, have roots in lived experiences that are space-time dependent, but they should be viewed as such. However, it is hard to disentangle past experience from future expectations and such space-time dependent negative interactions become border markers and cautionary tales which, for some, limit movement and use of the city space. Therefore, unpacking how intractable conflict narratives are established is an important

© The Author(s) 2019   1
S. Forde, *Movement as Conflict Transformation*, Rethinking Peace and Conflict Studies, https://doi.org/10.1007/978-3-319-92660-5_1

part of establishing ideal conditions for conflict transformation. Then the question is, how do cities become divided? The answer is typically due to political and institutional divisions materialised through urban planning and jurisdictional divisions which can direct the social use of space. Therefore, narratives of social movement are important to consult in order to understand the extent to which institutional divisions impact on the use of city space, and to explore how other social divisions direct movement in the city. Space and time are interrelated, we know this at a basic level because in our everyday lives we are generally supposed to be in a certain space at a certain time, we order our lives in space with respect to time. In this work, the interrelation is demonstrated through the transformation of spaces, and memory, through movement over time. Space, time and memory work together and contribute to the formation of spatial identities, these are not static constructs, but are malleable and, importantly, up for negotiation- and transformation.

THE CITY OF MOSTAR

Preceding the 1992–1995 conflict, Mostar was a popular tourist location, drawing visitors to the namesake bridge, Stari Most (Old Bridge). The name, Mostar, is a derivative of mostari or 'bridge keepers' (UNESCO 2005: 178) which demonstrates the importance of the bridge in Mostarian identity and heritage. The destruction of the bridge in 1993 became symbolic of the divisiveness of the Bosnian war and the far-reaching consequences of the violence, which transcended the space, but also the period of the war. While the reconstruction of the bridge was hailed as providing reconciliation in the city this can be regarded as only having been realised symbolically as the city remains divided (Calame and Pašić 2009). Over the course of the 1992–1995 Bosnian war the city was heavily damaged, by the end of the war the population of Mostar was largely displaced, and the divide in the city was entrenched. This was enshrined by the Dayton Agreement which cemented the conflict divisions nation-wide. Mostar embodies the wider country, as divisions in both the city and the country are spatialised. The divide in Mostar is maintained across a spectrum of actors operating across different spaces and is intermeshed with the formation of identity (Björkdahl 2015: 113). While spatial divisions within cities are not limited to post-conflict contexts, intrastate conflict can exacerbate other divisions (class, gender, sexual orientation, age) along the fault lines of ethno-nationalistic identities. In post-conflict academic discourse, the city of Mostar is characterised by the divided city narrative, this is not to say that the city is not divided, however, there is a risk of condensing diverse narratives of space which can overlook conflict transformation at a local level. Therefore, as Palmberger (2013: 558) proposes, consulting 'personal accounts' in Mostar can explore cross boundary movement and previous or current shared spaces of movement. Personal narratives also provide two important elements to research, firstly, they provide an on the ground qualitative account of the use of the city space which provides a different understanding of the space. Secondly, through these narratives, the agentive capability of social actors to transform space can be demonstrated.

The 1992–1995 war occurred following the dissolution of the conceptualisation of the former Yugoslavia. In institutional narratives and journalistic reporting, binaries emerged through the categorisation of ethnicities involved in the conflict. The designated identities of Bosniak, Serb, and Croat in political, academic, and social narratives became short hand for an assumed lineage or heritage and fundamentally denote a shared experience of space. This book unpacks place experience from these binaries, but for functional use the designations of 'Bosniak', 'Serb', and 'Croat' are used. These designations are used in acknowledgement that there is no absolute experience drawn from these identity categorisations in past, present or future social spaces. Therefore, while these identities are important, there is potential for these designations to be applied divisively. It is important to discuss the signifiers of the 'Bosniak', 'Serb', and 'Croat' labels as those that permeated the conflict and divided former neighbours and friends, but to underscore that designations of ethnic identity did not result in a shared experience. In summary, categorisations of individuals as 'Bosniak', 'Serb', and Croat' does not represent communal experiences inclusive of these ethnic designations, as complex aspects of identity and social roles can transcend these binaries as variables of social movement.

THE SOCIAL NARRATIVES

In June 2014, I first went to Mostar to discuss the potential research approach with local residents. During this first trip I proposed the use of maps as the methodology of the research. The maps were to be used to spatialise movement in order to explore the extent to which

institutional narratives of divisions direct social movement. The methodology was positively received and the following year across March and
April 2015, the first round of data collection took place in the form of
individual interviews with participants, in these interviews participants
drew and narrated a map of their social movement. Following on from
this trip in September 2015 the Map of Mostar workshop was conducted
at Abrašević. The workshop turned into a group interview with a group
map produced, further interviews and follow-up interviews were also
conducted, with some follow up questions also conducted over email.
During field research individuals were approached to participate through
different channels. This took the form of emailing, visiting the student
unions of the universities in the city and talking to residents. The following is a small summary of how participants became involved in the
research process. In April 2015 I emailed the Rock School director,
after learning about the school in the city. The director, participant S,
agreed to talk the next day, and after an informal conversation about
the research he introduced me to two students at the school waiting to
start their music lesson, who became participants A and B. Participant
C (a member of upper management at UWC) and N (charity worker)
were first contacted via email, after the interview participant C then
introduced me to two UWC students' D and E. While participant N
introduced me to their colleagues O, P and Q, who all took part in a
group interview. I met participant K (current youth politician and former Student Union President) at the University Džemal Bijedić Student
Union office; and he put me in contact with participant M (former
Student Union president of The University of Mostar and founder of
an NGO in the city). I first met with participant J (former director of
OKC Abrašević) for an informal chat after emailing the centre. In April
2015 participant J then put me in touch with participant F (international volunteer), G (one of the Street Arts Festival organisers), and in
September 2015 put me in touch with participant H (former director
of OKC Abrašević). I came into contact with participant L (from the
Mostar planning department) as they had been speaking at a workshop
in the city. Similarly, I met participant I while walking around Mostar,
they also took part in the workshop and put me in touch with participant R (from the Sarajevo planning department). A letter was assigned
to each participant to ensure anonymity and corresponds with the order
in which the interviews or exchanges took place. The randomised and
opportunity-based recruitment of participants allowed for interest driven

involvement with the research. The above participants' narratives of spaces and places in the city are the focus of this book.

SPACE AND PLACE

Experiences of space could be interpreted as a conflict; new experiences cause one spatial reality to cease to exist through the creation of a new one. Understanding this as a process of transformation can help analysis at all levels in order to understand the interconnectivity of spatial agency. We may feel this tension when returning somewhere after a long time away, what was once familiar is somewhat unfamiliar though it is the same 'place' it is no longer the same space for us. Fundamentally, though we are the main narrator of our experiences of space, spaces are social, our experiences of them is informed by those we interact with within them. Places, are a collective accumulation of social narratives of spaces, and represent spatial power dynamics. If we think about a internationally well-known city such as London we do not first think of the terraced houses in Clapham or the graffiti on an underpass. What locations come to mind are those that are, to borrow a term from physics, blackholes of social movement in that they have an intense social "gravity" which draws further movement. Many such spaces are symbolic of political and economic power structures which dominate our everyday lives. These spaces establish a collective understanding of the history of place. From grand boulevards to high-rise office blocks, and gated communities, the urban environment is readable of who, or what system, holds power. These spaces are designed to encourage, and equally, to dissuade the movement of particular people. Places, like intrastate conflict, are typically directed from the top-down this means that they are stages of domination and power.

As experiences of space are personal, our conceptions of space rarely align precisely, though some spaces are more divisive than others. While there is a fluidity in the narrative of spaces, they are not abstracted from power dynamics, as Foucault observed 'we live inside a set of relations that delineates sites which are irreducible to one another and absolutely not superimposable on one another' (1984: 3). So, while our experience of space is individual and socially informed, the production of physical space in the context of cities, is a top-down directed process. Therefore, as conflict is directed from the top-down, the post-conflict reconstruction of cities can become a continued spatial narrative of the conflict as

political actors seek to enshrine any territorial gains. Though the production of space—what this research proposes as the staging of space—is
directed from the top down, social actors can, and do, use the space in
an alternative function—what this research proposes as the scripting of
space. Though the institutional ethno-nationalistic divisions in the city
of Mostar and the country of Bosnia-Herzegovina (herein referred to as
BiH, when suitable) maintain the narrative of the conflict, social actors
transgress this divisional staging and rescript the post-conflict space,
therefore facilitating (though perhaps sometimes unintentionally) conflict transformation.

SPACE-TIME

The word space is used throughout the book to refer to physical but also
conceptual spaces. Space, is not static but is always transforming through
time. Massey's (1994: 260) observations regarding space-time relations
borrows terminology from the physical sciences and Einstein's general
theory of relativity. Essentially, the three dimensions of space (up-down,
forward-backward, left-right) must be thought of and analysed in combination with the fourth dimension of time (Massey 1994: 261). As we all
move along our own complex "timelines" of life we observe others moving through space and time, this also has an impact on our use of different spaces. There are notably, multiple spaces (activated by social use)
in physical places, which may or may not be seen or in use at all times.
Reflecting on space-time relativity, Massey (1994: 3–4) sets out that 'the
observer is inevitably within the world (the space) being observed' and
that there is a 'multiplicity of different spaces or takes on space' as each
social actor moves 'relative to one another, each thinking of themselves
at rest'. Inevitably, there is a 'dislocation within space/space-time' as,
individually, people are 'conceptualising and acting on different spatialities' (Massey 1994: 4). Essentially, we all have a particular spatial trajectory which shapes our experience of, and interaction with others in
space. Fundamentally, the social experience of space does not correlate
explicitly person to person, there are different external stimuli which
have influenced and continue to influence individual use and perceptions of space. This includes ourselves, though we may not realise the
impact our presence may have on others. That is not to say that there are
no overlaps in the understanding of space, indeed there is often a social
consensus on the use of spaces, as Massey (1994: 4) notes the 'spatial

organisation of society' is fundamental in the creation of 'the social'. As space is a social product (the constant production of which will be explored further on) it is regarded as existing at all levels (Lefebvre 2009) from the global to the local. These levels are enmeshed via 'social interrelations and interactions at all spatial scales' (Massey 1994: 264). Space is essentially a 'moment' composed of complex, layered and inter-secting social relations (Massey 1994: 265; Tuan 1977: 17). The word moment, is useful but also appropriate as it is temporally weighted, and our experience of space is ever-changing as the social actor, or observer interacts/views the space. This is something that constantly changes, due to other intersecting social experiences of space, and the presence of others. This includes the complexity of not simply what that other individual is doing, but what we expect them to do. This is often directed by stereotyping of 'ascribed identities' and in divided cities can spatially perpetuate the transgenerational conflict divides (Newman 2013: 27). Therefore, when thinking about space it must be remembered that in usage, space does not refer to an 'absence or lack' but that which is complex dynamic, fluid, evolving, and informed by our understanding or sensing (Massey 1994: 261). The interactivity of space-time will be further explored in the coming chapters of this work, but for now, we can set out that space is imbued with social interactions which are trans-formed throughout time.

STAGING AND SCRIPTING SPACE AND PLACE

While there are ethno-nationalistic divisions in the city of Mostar. As previously noted, focusing exclusively on this division discounts sub-divisions which may exacerbate or traverse the conflict division. Other variables such as age, gender, and even personal tastes, which impact on use of city space are explored through participant narratives of movement. The narratives introduced throughout later chapters demonstrate the potential for movement to transform divisive space into shared space in divided cities. The spaces which I apply the descriptor 'shared' to, operate without ethno-nationalistic identity signifiers and offer activities, and opportunities for shared social movement (Gaffikin and Morrissey 2011: 102). Critically, historical and contemporary social relations in space inform expected future interactions which impacts on how space is used. This is directed by complex individual and social understandings of narratives of place and space. Accordingly, the spatial impact of new interactions with different social actors in new social spaces is not limited to that physical place.

For participants, the locations which operated without ethno-nationalistic signifiers or divisions evoked the most appeal. The evident popularity of such spaces, outside of the participants' usage, indicates the importance of the establishment and facilitation of spaces which operate across the ethno-nationalistic conflict divide (Ethnographic research 2015). In particular, this research regards Omladinski Kulturni Centar (Youth Cultural Centre) Abrašević (herein OKC Abrašević or Abrašević), as a unique shared space in the city in that it is locally established and led, it hosts a variety of events which enhance the capabilities of social actors to transform the city. In establishment and ongoing operation, the centre received, and receives international financial support while maintaining autonomy in its activities. This demonstrates the capability for local actors to rescript and restage space when supported institutionally. However, Abrašević is only one example of a space in the city which has been staged in one function, derelict from the conflict, and has been socially transformed to operate in a different function.

While this work finds that social actors are agentive and can transform meaning and social relations in spaces, the influence of capital and institutional actors is fundamental in the construction or staging of place. The influence of capital and institutional actors is presented as staging and restaging. Staging does not occur in isolation from social actors because individuals facilitate the functioning of institutions and lend legitimacy to institutions. However, residents of a city are not typically included in the process of city planning, though such individuals are dominant users of the urban space. Although unable to independently restage the city, residents have the ability to spatially resist structured cityscapes through their social movements which rescripts space. The following "equations", propose the typical interaction between actors, which results in the rescripting or restaging of a space (Fig 1.1).

The (re-)scripting of space is set out as a bottom-up process enacted by social and transgenerational actors, while the process of (re-)staging is a top-down process enacted by institutional actors and facilitated by capital as economic means. In acknowledgement of this, the conceptualisation of social scripting, and rescripting refers to the ability of social and transgenerational actors, as residents, to transgress or reaffirm institutional divisions through movement. There is a duality in what I define as social rescripting which refers to Goffman's (1971) dramaturgical theory

Fig. 1.1 Rescripting and restaging space

Social actors + transgenerational actors = (re-)scripting

Institutional actors + capital = (re-)staging

and the social performance of 'everyday life'. The performativity of social interactions, as theorised by Goffman (1971) through dramaturgy, is spatially based. This process involves social actors tailoring their performance to different social stages and social audiences. In our everyday lives we upkeep multiple performances and hold different reoccurring roles, in essence we have different scripts and therefore projected personas for different social stages. As socio-spatial beings we have historically established and continue to build on and use 'place' as a point of physical anchorage, and a point of self-identification. We create, transform and are in turn transformed by space, it is a fluid, multifaceted relationship, one that is dually social and individual. In this work, the "actor" is theorised as an "agent" capable of transforming the stage through socially 'scripting' physical spaces (or stages of social performance) via their movement and use of space. Ultimately, (conflict) transformation is possible in the rescripting of place which can transform institutional staging of space (Lederach 1995: 7). To further contextualise this, much in the same way that a stage is set for a play or performance, the staging of cityscapes differs dependent on the intended use of the space and the expected social performance of that space. For example, a park is staged with a different planned use to a shopping mall. Though a shopping mall is staged as a place of consumerism, it has the potential to be rescripted with the comparative functions of a park, as a place to meet with others without buying anything. The socialness of consumer spaces is of course key to the design of modern shopping centres. This is what Bryman terms 'hybrid consumption', the overall aim of these spaces is to increase the time spent in them in order to increase the money spent in them and to establish the shopping centre 'as destination' (2004: 57–58). Furthermore, frequently 'private' commercial spaces are presented as public thoroughfares, as increasingly public space is privatised but maintains the illusion of it being open to the public, creating 'pseudo-public spaces' of consumerism under the guise of public space (Shenker 2017). These spaces can be used without spending any money, though they remain spaces of surveillance, power, and control, to keep certain groups, such as 'the homeless…[or] groups of rowdy teenagers' from the space

(Bryman 2004: 145). Consumer spaces are therefore, not always as open as they seem. Nevertheless, social use of space can correlate, clash or mix with the institutional staging. This is interactive with other actors who use the same space and has the potential to rescript space, engage with rescripting performed by others, and transform not only some narratives of space, but social relations within space. The two are interdependent as space, and narratives of space are produced by social relations, which are transformed by narratives of space. As observed by Skrzypek (2013: 9), in social spaces 'the audience has an impact on how the performance develops and how it ends, it is at least as important as the actors themselves.' This is what we can also take away from Massey's (1994: 260) theorisation of space-time, as we are all moving relative to one another, the trajectories we come from, and our interactions along the way jettison us along certain socio-spatial routes which affects our perceptions of space, relative to others. Accordingly, in all spaces, actors and audience are interchangeable in the rescripting of city space. Through this interchangeable interactivity, spaces can be rescripted as shared, and divided social relations can be transformed. In summation, scripting is conceptualised as the 'symbolic transformation of a space through social usage' and is a socially led process, whereby social use of a physical space is affirmed by actors who use that space (Forde 2016: 468). Whereas, staging as the 'physical transformation of a space through deliberate damage or restoration' is a top-down led process through which actors set out a defined use of a location through physical construction (Forde 2016: 468). All physical transformation thereafter would be described as restaging, the same applies to rescripting. The spatio-historic terms are employed to acknowledge the indelibility of socio-spatial interactions in the ever transforming or contested nature or identity of place.

The World in Motion

If there are no fixed points then where is here? (Massey 2005: 139)

The above quote is a summative remark of Massey's geographic analysis of the unseen movement of a mountain, used as an example to demonstrate the inescapable transitionary "movement" which all living beings and physical things experience. From the erosion of a mountain to the grey hairs on our head and fine lines around our eyes, we are all moved by time. To further illustrate this, Massey (2005: 139) looks to the stars, noting that a seemingly stable part of our navigation, the northern pole star, was not always the north star. As Massey (2005: 139) observes, this might be logical, but critically Massey emphasises that, 'the point is to feel it, to live in its imagination', this is necessary to truly grasp the transient nature of space as that which is inherently tied to movement and transformation. Accordingly, we must aim to expand our understanding of space and how our spatial trajectory informs our analysis of space. In the example of a mountain eroding or the tide causing landmass movement, the speed with which we experience time impacts on our perceptions of change in space (Massey 2005: 139). As we speed through our lifetimes it may not seem like a mountain has moved much and therefore it seems static, though it is always moving or transforming. This is essentially, space-time relativity, as we move past or through one event our perception of it changes dependent on the speed at which we are travelling. This is an important lesson for post-conflict researchers, for if we are to understand complex post-conflict spatial movements and relations, we must spend some time exploring them. As Kaplan (2012: xii) observes 'a good place to understand the present, and to ask questions about the future, is on the ground, travelling as slowly as possible'. That is to say, small changes over time are inevitably overlooked by researchers, either "travelling too fast", or lacking the "imagination" to conceive that social actors have the agency to change space in subtle ways which can have large-scale and long term effects.

Space is frequently or traditionally understood to be that which is conceptual, empty or void. Whereas place, filled with meaning and identity, is brimming with identity and social relations. All place can be viewed as space. However, space, has a conceptual advantage due to multilayered definitions and experiences therein. Comparatively, understandings of place are somewhat less flexible, as Tuan (1977: 179) suggests, place is a static concept as 'an organised world of meaning'. However, place, and place identity, are constantly being rescripted and edited over time. We put things in place, we move in space. Borrowing from the physical sciences once more, space is infinite, place on the other hand is, well, placed. Understandings of place and space are experienced momentarily. The spatio-temporal nature of our understandings of space and time, our experiences of the here, are accumulated and 'build up a history' of place, (Massey 2005: 139) when someone revisits a space it will have transformed in some way. Fundamentally, cities exist in multiple space-times, they exist as a living urban centre, in memories, books, tour

guides, postcards, and movies. They are spaces of prosperity and afflu-
ence, and extreme poverty and deprivation. Each city is an ever evolving
spatialised narrative of those who live in and move through the physical
space. Every day the spatial narrative of any city is rescripted by residents
who move through and use the space. The narrative may be physicalised,
in the streets and the structure of the city or may be may exist in stories
and narratives of space. An individual will not experience a city in the
same way as another person might, but due to 'ascribed' and 'achieved'
identities, commonalities of experiences of movement are identifiable
(Newman, 2013: 27). These experiences generally exist in a defined
space[1] as the construction of city space is directed or staged from the
top-down, as Massey (2005: 166) notes the;

> power-geometry of space time [...] is one of power and politics as
> refracted through and often actively manipulating space and place, not one
> of general 'rules' of space and place.

In this analysis, Massey (2005: 166) directs us to think about the power
relations in space, which are constantly shifting space-time experiences.
These may shift slowly, for example, in Mostar and other post-conflict
divided cities, divisions remaining from the conflict are still identifiable.
In Mostar, the political divisions are enshrined in jurisdictional and educational divisions, which attempt to direct movement. This maintains a
narrative of the ownership of space, materialised through the divisiveness
of the extreme violence committed during the war. In other divided cities, the spatial barriers are established in similar but also different ways.
In Belfast, peace walls and a divided education system separates communities and maintains the divisions of conflict in Northern Ireland.
Comparably, in Cape Town, and South Africa at large, the forced remov-
als and cultural cleansing of the institutionalised racism of apartheid
(Afrikaans for the policy of racial segregation) displaced and disintegrated
communities into townships. The displacement was threefold and spatially, politically, and economically marginalised communities. The spatial
displacement impacts on what spaces individuals use or have access to, so
while education in South Africa is officially open it is very much divided
in terms of access. Furthermore, those who live in townships are more
likely to have to travel long distances for work and suffer crime due to
socio-economic marginalisation.

Space is power. In divided cities such as Mostar, Belfast, and Cape Town the structurally implemented spatial divisions are a continuation of the conflict and create the conditions for a transgenerational transference of the conflict. They are distinctly spatial in nature and are formed through, and maintained, by boundaries which direct movement, use, and access to space. All space tells a story, the 'power-geometry' (Massey 2005: 166) staging of Belfast and Mostar demonstrates that the spatialised divides of political understandings of ownerships of space has established the existence of divided communities. In these two cities the divide is negotiated on a smaller spatial scale with the parties contesting the space itself. The 'power-geometry' staging of Cape Town, a significantly larger city, demonstrates the use of space and displacement as the narrative peripheralisation and a continued act of violence (Massey 2005: 166). The maintained divides in the city marginalise the black, coloured, and immigrant communities in townships, while the socio-economic gulf reinforced by spatial dislocation, maintained by gentrification, preserves an apartheid in the city.

While cities are staged top-down in their structure, memory, and that which is shared and becomes social memory,[2] is scripted in space through the social use of the space, and in some cases through socially directed constructions in the city space. This is effected by the condition of the staged space and is important to consider following an intra-state conflict due to the impact this type of conflict has on local infrastructure. For example, in Mostar buildings used prior to the war, now derelict due to damage from the war, are maintained as a stage of the 1992–1995 conflict through their unavailability for social movement. Though these spaces are still used by some, in ways unintended, generally they are spaces of non-movement. An important element for conflict transformation is facilitating movement and use of space, this involves the restoration or renovation of properties damaged or destroyed during the conflict. This should be regarded to be a priority in post-conflict spaces. However, it is a process that has been visibly stalled Mostar and exacerbated as unfinished construction projects join the derelict buildings. The issue of stalled post-conflict reconstruction in the city space, visualises the precarious nature of the transformation of post-conflict city-space and maintains the memory of the conflict through the physicality of the city space. Spatial social memory may be established through every-day space or through spaces formalised 'stages of memory' whereby a specific institutional narrative is set out (Forde 2016: 468). Though this would typically refer to formal spaces or memorials the continued visibility of space damaged during the war preserves the city as a stage of memory of the conflict. Critically, the lack of post-conflict reconstruction in the city has the potential to perpetuate post-conflict memories, disenchantment with the government, and overall, restricts social movement and the transformation of space. However, while there are institutional 'glass ceilings' in spatial transformation, narratives of space are in constant transformation and negotiation by social actors.

RESEARCHING SPACE AND CONFLICT

In cities divided spatially, the social use of space is a measure of the social negotiation of conflict divisions. However, there are certain assumptions regarding post-conflict spaces, as observed by Björkdahl and Höglund (2013: 291) that 'post-conflict spaces' are understood by external institutional actors to be 'empty spaces in need of new norms, practices and governing institutions'. This international persepctive can be considered to be an orientalist lens (or in this case Balkanist lens) which locates dissimilarities as a deficit of stability, and in turn delegitimises local involvement (Said 1979; Todorova 2009: 3; Björkdahl and Höglund 2013: 291). The perception of post conflict space as empty has been traditionally remedied by the application of external liberal processes. However, through efforts to fill the perceived "emptiness" of post-conflict spaces, liberal definitions of peace can be observed as addressing the 'episode (visible expression of conflict)' but failing to unpack and consider the urban, social, and personal contexts of conflict, or what Lederach (2003: 31) regards to be the 'epicentre of the conflict'. Through conflict transformation, there is potential for a process of closure wherein 'the loss is to a large degree accepted and incorporated into the functioning of everyday life' (Hamber and Wilson 2002: 37). In divided cities, everyday life or everyday use of space is frequently directed by the conflict divisions, however, a space directed by institutional definitions of space may be reclaimed or challenged through movement which traverses the divisions. Fundamentally, our understanding of space is heavily influenced by what we "know", as Soja (1989: 120) notes, 'spatiality cannot be completely separated from physical and psychological spaces'. The relationship between psychological space and physical space is considerable, as the perception of 'the spatial separation of places itself' directly influences behaviour or the use of space (Briggs 2005: 363). This can be affected by emotive value or understandings of space which influences an individual's understanding of the distance of one place from another (O'Keef and Nadel 1978: 76–77). For example, though there are no borders to cross in Mostar, and no walls which divide the city, boundaries have been fostered due to institutional divisions which directed the conflict in the city, trickling down to impact on social use of the city space. However, it may be argued that the process of exploring, and processing, 'new' space through social movement can transform perceptions of space.

The study of space and movement is rooted in behavioural and human geography which encompass academic inquiry of how people use space. As such, it is concerned with human action within an environment 'according to [an individual's] perception and understanding of the situation' (Castree et al. 2013: 30). Essentially this concerns how and why individuals move through space. The origins of behavioural geography stem from Tolman's work (1948) 'Cognitive Maps in Rats and Men'. Cognitive maps are set out by Tolman (1948) as a 'mental representation of the spatial layout' which concerns how we process space (Eichenbaum and Cohen 2004: 50). Tolman (1948: 189–208) theorised two types of cognitive maps, 'narrow strip' and 'broader comprehensive', and theoretically linked them to human behaviour. Behavioural reflexes such as 'regression, fixation and displacement of aggression onto out-groups' are considered by Tolman to exemplify a 'narrow strip variety' of a cognitive map (1948: 189–208). The importance of peace or 'calm' and resources or being 'well-fed' in a social environment is presented by Tolman as essential to develop 'truly comprehensive maps' (Tolman 1948: 189–208). In other words, when core needs are not met, individuals have a narrower range of spatial perception, which can result in the negative behavioural impulses. However, when an individual's needs are satisfied their social map can broaden. This demonstrates the importance of stability, peace and resources to foster conditions in which individuals may engage with space, and other social actors within space. The link between perceptions of space and actual space is further explored by O'Keefe and Nadel (1978: 6–7) who developed Tolman's work and observed the interdependence of psychological and physical spaces. Specifically, O'Keefe and Nadel (1978: 6–7) outline the influence of physical space on psychological space. Fundamentally, this demonstrates the way in which experiences of staged space may impact on perceptions of space and therefore influences (limits or encourages) social scripting or use of space. Therefore, understanding the social use of space in divided cities can help establish the extent to which institutional divisions trickle down and limit movement but how individuals also 'create their own historical myths' (Tuan 1976: 266). Human geography is complimentary to critical peace studies, or bottom-up peacebuilding as it recognises the agentive capabilities of social actors to produce and transform narratives of space. Additionally, the psychological and social construction of spaces is also demonstrated as human geography observes place as 'a focus of human emotional attachment' (Entrikin 1976: 616). The above discussion of the socio-psychological construction of perceptions of space evidences the interactivity of institutional definitions of place and their impact on social use of space, while human geography allows for an exploration of the social nature of space and place.

SOCIAL MAPS

In post-conflict divided spaces, institutional narratives of division and fearmongering of the have the potential to limit social movement. However, the spatial problem can also become the spatial solution Therefore, in post-conflict spaces and deeply divided cities, local perceptions of space can help researchers understand socio-spatial transformation and can help practitioners and institutional actors understand social needs. Maps are a traditional tool for understanding space, directing, and exploring movement in space. Historically they have been cartographies of power and domination establishing a top-down narrative of space. However, as an object of power they can be reclaimed, and socially generated maps can challenge, and contest established power structures. Accordingly, mapping was employed as a methodology to explore the spatial agency of participants and to evidence the interaction between the personal/social, urban, and institutional[3] spaces in which peace is fostered. Socially generated maps visualise the way individuals utilise space every-day. This movement is of course public but also intrinsically private as it is performed in public space, but the narrative, feeling, or purpose of movement is generally a personal experience. Such spatially informed research approaches have been utilised in other divided cities such as Belfast (Murphy and Murtagh, 2010) and Jerusalem (Raanan and Shoval 2014). These typically private narratives can evidence the sensations associated with space, as it scripts and rescripts 'overlapping worlds' of personal interaction; within the wider context of the social and divided urban and institutional spaces (Mitchell 2011: 7). Meaning attributed to space through social movement -which may or may not correspond with institutional definitions of space- is what this research presents as rescripting. Individuals who are involved in rescripting of space so that it becomes open or shared, are proposed as agentive and actively involved in a conflict transformation processes (Lederach 1995: 7). Additionally, mapping has potential to indicate subtle or unintentional conflict transformation through social movement. Therefore, exploring the narrative of spatial movement in divided cities can help unpack the conflict narratives and discuss the potential transformation of the space.

The narratives in this book were produced as part of narrative based mapping activities. This involved a participant either using a road map to mark out social spaces or drawing their own social map of Mostar. Hand drawn maps of spaces are physicalisations of our cognitive maps, formed by 'visual…and non-visual…means of information' (Haken and Portugali 2007: 59). Due to the social and sensory nature of our experiences of space, the expression of this individually compiled knowledge must also be multi-modal and so 'narrative interviewing' was employed alongside the mapping (Bogner and Rosenthal 2014: 9). It is notable that areas of non-movement (areas in which a participant would not usually go to or move within) are non-divisive in terms of functional movement (shops, work, university and home). However, as the institutional maintenance of ethnicised space can perpetuate conflict divisions spatially, and transgenerationally, discussing context of the movement (such as intentionality or unintentionality) can help establish if movement is directed by the conflict divide. Through these spatialised narratives, the extent to which social movement reflects or traverses institutional divisions; rescripts institutional divisions or indicates sub-divisions can be illustrated. Critically, one of the strengths of cognitive mapping is also a key weakness, in that the researcher is responsible for the interpretation of the data produced by the map. The responsibility of the researcher is of particular importance in the context of divided cities, as the researcher should be conscious not to co-opt narratives. For example, in Mostar, there is a notably higher number of social spaces on the 'Croat' side of the city. Therefore, if focusing exclusively on the location of spaces of movement, the narrative of the ethnic division could become the dominant narrative of the movement, to the exclusion of other potential divisions in the city. What the researcher then must do is to take a closer look at narratives of movement in divided cities. As this chapter has discussed, there are multiple narratives in any one space and these narratives do not always correspond with institutional definitions of space. Space is flexible, transitionary, and momentary and therefore, there will always be opportunities for rescripting and transformation.

NOTES

1. However, this does not just apply to urban space as even rural spaces have borders which may be constructed or natural, for example, agricultural land, rivers, thick forests, and canyons may all restrict movement.
2. All memory can be considered social, in that it is formed through or in relation to other individuals or objects made by others.
3. This is explored in-depth in chapter three.

BIBLIOGRAPHY

Björkdahl, A. (2015). 'Two Schools Under One Roof' Unification in the Divided City of Mostar. In A. Björkdahl & L. Strömbom (Eds.), *Divided Cities, Governing Diversity*. Lund: Nordic Academic Press.

Björkdahl, A., & Höglund, K. (2013). Precarious Peacebuilding: Friction in Global–Local Encounters. *Peacebuilding, 1*(3), 289–299.

Bogner, A., & Rosenthal, G. (2014). The "Untold" Stories of Outsiders and Their Significance for the Analysis of (Post-) Conflict Figurations. Interviews with Victims of Collective Violence in Northern Uganda (West Nile). *Forum: Qualitative Social Researching. Sozialforschung, 15*(3). Art. 4. Available from: http://www.qualitative-research.net/index.php/fqs/article/view/2138/3705. Accessed 10 Jan 2015.

Briggs, R. (2005). Urban Cognitive Distances. In R. Downs. & D. Stea (Eds.), *Image and Environment: Cognitive Mapping and Spatial Behaviour*. New Brunswick, NJ: Transaction Publishers.

Bryman, A. (2004). *Disneytisation*. London: Sage.

Calame, J., & Pašić, A. (2009). *Post-conflict Reconstruction in Mostar: Cart Before the Horse. Divided Cities/Contested States* (Working Paper. No. 7). Available from: http://www.conflictincities.org/PDFs/WorkingPaper7_26.3.09.pdf. Accessed 20 June 2014.

Castree, N., Kitchin, R., & Rogers, A. (2013). Dictionary of Human Geography. Oxford: Oxford University Press.

Eichenbaum, H., & Cohen, N. J. (2004). *From Conditioning to Conscious Recollection: Memory Systems of the Brain*. Oxford: Oxford University Press.

Entrikin, N. T. (1976, December). Contemporary Humanism in Geography. *Annals of the Association of American Geographers, 66*(4), 615–632.

Forde, S. (2016). The Bridge on the Neretva: Stari Most as a Stage of Memory in Post-conflict Mostar, Bosnia-Herzegovina. *Cooperation and Conflict,* 51(4), Available from: http://journals.sagepub.com/doi/pdf/10.1177/0010836716652430.

Foucault, M. (1984, October). *Of Other Spaces: Utopias and Heterotopias. Architecture /Mouvement/ Continuité* (J. Miskowiec, Trans.). Available from: http://web.mit.edu/allanmc/www/foucault1.pdf. Accessed 20 Jan 2017.

Gaffikin, F., & Morrissey, M. (2011). *Planning in Divided Cities.* Oxford: Wiley Blackwell.

Goffman, E. (1971). *The Presentation of Self in Everyday Life.* Harmondsworth: Penguin.

Haken, H., & Portugali, J. (2007/1996). Synergetics, Inter-representation Networks and Cognitive Maps. In J. Portugali (Ed.), *The Construction of Cognitive Maps.* Dordrecht: Kluwer Academic.

Hamber, B., & Wilson, R. A. (2002, August 3). Symbolic Closure Through Memory, Reparation and Revenge in Post-conflict Societies. *Journal of Human Rights,* 1(1), 35–53.

Kaplan, R. D. (2012). *The Revenge of Geography.* New York: Random House.

Lederach, J. P. (1995). *Preparing for Peace: Conflict Transformation Across Cultures.* Syracuse: Syracuse University Press.

Lederach, J. P. (2003). *The Little Book of Conflict Transformation.* Intercourse, PA: Good Books.

Lefebvre, H. (2009). *The Production of Space* (D. Nicholson-Smith, Trans.). Oxford: Blackwell.

Massey, D. (1994). *Space, Place and Gender.* Cambridge: Wiley.

Massey, D. (2005). *For space.* London: Sage.

Mitchell, A. (2011). *Lost in Transformation: Violent Peace and Peaceful Conflict in Northern Ireland. Re-Thinking Peace and Conflict Studies.* Basingstoke: Palgrave Macmillan.

Murphy, A., & Murtagh B. (2010). *Children, Policy and the Built Environment* (Working Paper No. 1). Queen's University Belfast. Available from: http://www.qub.ac.uk/research-centres/TheInstituteofSpatialandEnvironmentalPlanning/FileStore/Filetoupload,199267,en.pdf. Accessed 24 Nov 2013.

Newman, D. M. (2013). *Sociology: Exploring the Architecture of Everyday Life.* Los Angeles: Sage.

O'Keefe, J., & Nadel, L. (1978). *The Hippocampus as a Cognitive Map.* Oxford: Oxford University Press.

Palmberger, M. (2013). Practices of Border Crossing in Post-war Bosnia and Herzegovina: The Case of Mostar. *Identities: Global Studies in Culture and Power.* Available from: http://www.tandfonline.com/loi/gide20. Accessed 20 Oct 2016.

Raanan, M. G., & Shoval, N. (2014). Mental Maps Compared to Actual Spatial Behaviour Using GPS Data a New Method for Investigating Segregation in Cities. *Cities, 36,* 28–40.

Said, E. W. (1979). *Orientalism.* New York: Random House.

Shenker, J. (2017, September 26). 'It's Really Shocking': UK Cities Refusing to Reveal Extent of Pseudo-public Space. *The Guardian.* Available from: https://www.theguardian.com/cities/2017/sep/26/its-really-shocking-uk-cities-refusing-to-reveal-extent-of-pseudo-public-space. Accessed Jan 2018.

Skrzypek, J. K. (2013). *Is Terrorism Theatre? Dramaturgical Metaphor in the Cases of Budyonnovsk, Dubrovka and Beslan.* Excerpt of Ph.D. thesis, University of St Andrews. Provided by Janina Karolina Skrzypek.

Soja, E. W. (1989). *Postmodern Geographies. The Reassertion of Space in Critical Social Theory.* London: Verso.

Todorova, M. (2009). *Imagining the Balkans.* Oxford, NY: Oxford University Press.

Tolman, E. C. (1948). Cognitive Maps in Rats and Men. *The Psychological Review, 55*(4), 189–208. Available from: http://psychclassics.yorku.ca/Tolman/Maps/maps.htm. Accessed 20 Nov 2013.

Tuan, Y. F. (1976, June). Humanistic Geography. *Annals of the Association of American Geographers, 66*(2), 226–276.

Tuan, Y. F. (1977). *Space and Place: The Perspective of Experience.* Minneapolis: University of Minnesota Press.

UNESCO. (2005, July 15). The Old Bridge Area of the City of Mostar. *World Heritage Scanned Nomination.* Available at: http://whc.unesco.org/uploads/nominations/946rev.pdf. Accessed 14 June 2014.

CHAPTER 2

Rescripting and Restaging: *Spatialising Structure and Agency*

Cities are made through movement and social use of space which leaves, sometimes undetectable, but undulating stories of the space. Through their heterogeneity, cities are by nature divisive; class, gender, ethnicity, and religion intersect, and dependent on our spatial trajectories we may experience spatial belonging or may feel "out of place" (in any space). Fundamentally, urban spaces are *'relational'* and constructed from shifting plates of 'jostling, potentially conflicting, trajectories' (Massey 2007: 89) In Mostar, multiple colonial occupations shaped the city scape, and institutional conceptualisations have drawn the space into multiple conflicts and now maintain the divisions of the 1992–1995 conflict. However, social actors give life to the city and engage in place making through interactions and movement in space. This is observed as a mutual relationship, as Kappler notes, although agency can 'transcend structural boundaries,' it 'cannot be viewed as isolated from its surrounding structures given it is always situational and contextual' (2014: 36). Indeed, certain structures provide opportunities for certain agentive capabilities to be exercised and the variables of our identity shape how we experience structural constraints. This of course is in consideration of aspects of identity such as gender, class, race, which also have a direct impact on the spaces we have access to. As Tonkiss (2005: 2) observes 'the organisation of space both provides the basis for social relations and offers a reflection of them.' Spaces are staged and scripted with narratives regarding who is meant to use the space, this reflects both power structures and social relations. Inherently, there is a tension between the

© The Author(s) 2019   21
S. Forde, *Movement as Conflict Transformation*, Rethinking Peace
and Conflict Studies, https://doi.org/10.1007/978-3-319-92660-5_2

production and use of space, in that 'urban spaces can be seen as structuring social relations and processes' these are in turn 'shaped by social action and meanings' (Tonkiss 2005: 2).

This chapter proposes a theoretical framework for looking in depth at this process including the construction, understanding, and transformation of space. In acknowledgement of the interconnectedness of the structure of space and the shaping of space by social use, this chapter discusses the spatialisation of agency and structure introduced in chapter one and theorises the capabilities of social actors to transform space. This refers to physical spaces of interaction and conceptual spaces, alongside spaces that are used temporally in an alternative capacity to which they are staged. The theoretical framework maps out the actors who direct space and those who use space; of course there are overlaps between positionalities, such as political actors using social space, and this can impact on 'private' agency of individuals outside of their 'public' role. The framework organises the analysis of space through the theoretical constructs of rescripting and restaging which are used discuss how relational actors influence the transformation of physical and conceptual spaces (personal, social, urban, and institutional). These four spaces, collapsed into three (personal-social, urban, and institutional) are regarded as (potential) spaces of peace, as the broad physical and conceptual spaces in which peace may be enacted. It should be noted that there are a multiplicity of spaces and places temporally operating within these spaces. The terms staging, and restaging are used to locate the structural capability of top-down actors to spatially transformation cityscapes. While, scripting and rescripting is used as a conceptualisation of the capabilities of bottom-up social agents to transform staging of space through movement. To recap, the former, restaging, is conceived as top-down processes performed by institutional actors and facilitated by capital; and the latter, rescripting, is defined as bottom-up processes performed by social and transgenerational actors. The word actor is used in this work as a dramaturgical construct wherein, the actor is agentive and capable of tailoring their social performance to certain socio-spatial stimuli. Goffman (1971: 34) discusses the social performance of the individual self, how our appearances and the spaces we use reflexively affect that performance. In situating analysis through a sociological lens, the work also evidences the importance of socialisation and an acknowledgement of the tensions between structure and agency, which effect the capabilities of the actor to act and react in certain situations.

We perform socially each day, the deftness and instinctiveness with which we alter our performances and our outer-selves is infinitely complex and is informed by moment to moment analysis, which is based on the information we have at hand (including our history of social relations, our socio-spatial trajectories or our positionalities). As Goffman (1971) notes in the preface of 'The Presentation of Self in Everyday Life' individuals tailor their social image and activities to the audience in receipt of the performance. For example, the vulnerability we display in front of loved ones, compared with authority and professionalism in a work context. When performances are not socially fitting, there occurs a 'performance disruption' Goffman (1971: 235–237) demonstrates that such disruptions can have far reaching consequences, personally, situationally, and in the wider social structure.[1] Goffman (1971: 237) observed that this framework for analysis, or understanding role disruption, may be different in other cultural settings. Indeed, the space of analysis is intrinsic to understanding and exploring the expected performance in space and therefore, how individuals transgress space and post-conflict roles. Performance breaks in the three spaces of the individual, the situation, and a wider social structure may not align with expected performances but may not be negative in nature at all. In particular, such breaks are intrinsically important in transforming not only post-conflict space, but any space at all. Placed in a divided city space, an individual may break with preconceived notions of "the other", by either moving in certain spaces or directly socialising across divides. This has the potential to change not only individual perceptions about one person, but can transform the space of interaction, and also over time, arguably the wider social structures which maintain the division.

Notably, performances in and with space, often creates divisions, though they may not be violent or negative in nature. As Tonkiss (2005: 30) observes, socially we are divisive and, fundamentally, 'people make sense of their world by connecting and separating things, by drawing distinctions and ordering relations, and these processes leave their mark in space'. This can of course, be a deliberate exercise, and may be as innocuous as creating a physical or social path to connect with others (Simmel 1997: 172) or may be through scriptings which may require interpretation and engagement and for the actor to understand the narrative of the space (Barthes 1997: 168). In essence, as we are socio-spatial beings, through movement and social interactions, we construct divisions of movement every day. In a divided city space, this may manifest as a spatial pattern of movement which then may encourage wider or increased movement or may be divisive in dissuading movement. However, this is not to romanticise or exaggerate any individual capability to perform such transformative interactions as in divided cities institutional structures may have fostered conditions in which agentive capabilities have been curtailed or directed. Therefore it is useful to reflect on the concept of agency, it is observable that 'the exercise of agency involves mutuality and interdependence as well as relations of domination and subordination' (Björkdahl and Selimovic 2015: 170). Essentially, social actors can be considered to have the capability to reinforce 'relations of inequality' but may also 'challenge power relations, [and] question existing norms and practices' (Björkdahl and Selimovic 2015: 170). However, the assessment of the agentive capabilities of individuals to challenge structures can be problematised when considering relations of structural inequalities. Our socio-spatial origins influence our agency, and structural constraints can severely limit our agentive capabilities to challenge structures, for example, an individual growing up in a township in Cape Town, or in a divided education system and a war-traumatised household are spatially dislocated, their movement is structured by institutional conceptualisations of space. Fundamentally, the circumstances of our births are not awarded equal opportunities to challenge power structures or exercise our agentive capabilities. Therein, there is a complexity to finding the seam between structure and agency, arguably the extent to which individuals can exercise agency will differ person to person with different positionalities afforded more agentive freedom than others.

We may take a key point away from this discussion of the agency, or action of an individual, in that it can counter or reaffirm the structure. Through social movement individuals can rescript space, and this rescripting may correlate or clash with institutional staging of space. While the top-down staging of space attempts to direct individuals within the cityscape, the ability of individuals to use space in an alternative function to its staging, demonstrates the agentive capabilities of social actors to transform space. As a form of spatial agency, social movement enacts rescripting within the institutionally divided staging of the city. In essence, a city may be politically divided but socially there may be more cross-boundary movement. Overtime, this may result in a change in electoral strategy and may lead to a breakdown in the maintenance of divisions. This is of course a slow process, and much like Massey's mountain metaphor discussed in chapter one, if we do not look close enough it may appear to be one that is absent. That said such a process is unlikely to be linear and without a conflict of narratives. It is therefore important to amplify transformative narratives of divided spaces, to make the unseen visible. As previously noted some divisions such as class, gender, and age may take precedent over ethno-nationalist divisions. Through an analysis of the wider context of social movement in the city through 'achieved' and 'unachieved identities', there is observably an eclectic and shifting patchwork of spaces of conflict and cohesion in the city (Newman 2013: 27).

Socio-Spatial Agency

The city—as a concept and physicality—is relevant not only as a centre of capital but dually as a social centre or social conglomerate in the physical world. While the term social has a variety of definitions, in this context, it is used to describe the collective use of the space by residents of the city. The city is a place in which the power of institutional actors is staged. In the context of Mostar, Björkdahl observes (2015: 117) institutional divides in Mostar as representative of 'spatial governmentality'. Through a spatialisation of Foucault's (1991: 102) concept of governmentality, Björkdahl (2015: 117) sets out that in Mostar, social actors 'regulate themselves and… become active participants in the process rather than objects of domination'. Social actors therefore can be considered to be performing the institutionally staged roles of division in the city, reaffirming the division socially. However, while there are actors who rescript the city as socially divided, and the potential for insecurity and social unrest is evident through this, so too is the potential for social cohesion and interaction within and across conflict divisions. In all spaces institutional actors (through the staging of the city) attempt, but do not always succeed, to direct the social performance of individuals and groups of individuals using the city space. As Kappler (2014: 22) observes agency can be indivually excercised but also emerges socially and as such can be fostered in constellations with other actors. Moreover, while agency operates 'in a fluid structural environment,' fundamentally, individuals can renegotiate and transform such structures (Kappler 2014: 23). This notably occurs in space, and as Björkdahl and Kappler (2017: 21) observe 'agency is situated in the construction of spatial meanings'. However, this is a temporal and sometimes fleeting process, if we think about different spaces in a city; for example, schools, music, leisure or sporting locations all bring together a variety of different people which can cross ethnic, class and gender identities. Though not intentional spaces for building peace or they have the potential to become spaces which transform social relations. This demonstrates the opportunity social movement provides for social actors to rescript space. Such capabilities may be time and context dependent and multiple scriptings may operate in one space as different actors will attribute different meaning to spaces.

In post-conflict literature, space, has until recently, been treated 'as the dead, the fixed, the undialectical, the immobile,' an observation Foucault made more generally about the study of space in 1980 (p. 177). Comparably post-conflict scholarship previously held this frozen conceptualisation of space to account through an ever-relevant presence of time. The term, post-conflict itself, defines a transition from conflict as a linear process. In this work, the term post-conflict is used to allude to the challenges faced by the city but with the awareness that the conflict is ongoing in different ways. Fundamentally, the inclusion of space as an analytical context for post-conflict studies is an ever-growing field of scholarship. As Gupta and Ferguson observed in 1992 (7) space had been;

> a kind of neutral grid on which cultural difference, historical memory, and societal organisation are inscribed. It is in this way that space functions as a central organising principle in the social sciences at the same time that it disappears from analytical purview.

As spatial beings, our consciousness operates through physical interaction, and space is socially created. It therefore follows that social actors can create and transform space though this transformation will vary in degrees of visibility and longevity. As Gupta and Ferguson (1992: 7) note social actors have the agentive capability to transform space or;

> to confound the established spatial orders, either through physical movement or through their own conceptual and political acts of re-imagination, [which] means that space and place can never be "given".

In divided cities, and in Mostar, physical movement is interlinked with non-physicalised spaces of ideology. Social actors transform narratives of space through movement and this has the potential to traverse conceptual 'ethnoscapes' and spatialised ideologies (Björkdahl 2015: 117).

The social use of space is important to discuss in post-conflict divided cities, wherein conflict divisions are entrenched thereby maintaining the narrative of the division. The availability of space, for social use, can be observed through movement in the space, which is often, as Kappler (2014: 30) notes, linked to 'visibility in the public sphere'. However, there is no single visible public sphere. As we travel along our own spatially lived trajectories, what is public and what is visible is critically observed differently person to person and is dependent on movement or use of space. Fundamentally, structure and agency are spatial, it is through movement that social actors contribute to a discourse of space which may align or overlap with the institutional staging of the space for a particular purpose.

Staging the City, Scripting the City

In the city of Mostar and the wider country of BiH, it is notable that the conflict divisions are maintained through jurisdictional divisions and supported by top-down actors to maintain political control. The urban environment is where divides (ethno-nationalistic, class, gender and transgenerational) are created, can be reinforced, but may also be transformed. The construction of social space occurs through (re-)staging (via capital and institutional actors and (re-)scripting (via transgenerational and social actors). The latter is informed by the social production of space or what Lefebvre (2009: 27) regards as the 'social character of space'. The concept of rescripting demonstrates the ability of social actors to change the staging of space and identifies the constant ebb and flow of social movement which impacts on the definition of place. Through movement and use, individuals socially define space, and edit the scripting of staged space with their movement. We move through and live in physical space. As a result, repeat movement contributes to the social memory or script of place. Though this may not be at all times engaged, visible, or at the forefront of our consciousness, it impacts on movement. As Lefebvre (2009: 229) notes 'nothing disappears completely', this is why the term rescripting is used as social movement is a constant narrative which impacts on the contemporary and future use of space, there is no one static script of place. Though the experience of place is personal, it is also social. However, the experience of place, though possible to be shared, can never be an explicit or absolute shared experience. For example, in the same city no two individuals will have the exact same spatial experience even when using the same space, this can include traversing conflict divides. It may be the same physical place, but due to individual spatial trajectories, the understanding of the space is from a different frame of reference, a different perspective, and through this a different spatial story is formed. Of course, as inherently complex social beings we may experience dually a freedom of movement at one time and a restriction on movement at another in the same place. There is an inherent flexibility to the rescripting of space through the multiplicity of variables which can impact on the movement and use of space which may be informed by past, current or expected usage. In cities, the visibility of signs and symbols are representative of a multiplicity of social actors with a potentially different spatial understanding of the space. While not always temporally present, these actors essentially visualise their narrative of movement and establish ownership of space. As a social product, physical places exist as a spatial construct of social knowledge which is established via settlement and movement through and within the space transgenerationally. This social imprint, or the scripting and rescripting of movement, which may be through interactions or simply signs, graffiti, or stickers, involves the transmission of information about the history and heritage of place. Through interaction, different social scripts reinforce distinct social rules of movement through, and within a place which establishes, changes, and ultimately transforms the place itself. This process is ongoing within the staged cityscape and as a result the staging or scripting of space is complex, echoing the wider structure versus agency debate. While the process is ongoing and one of constant transformation, there is a consensus on what makes place due to the institutional staging of place identity. It can be outlined that social movement is directed not only by the institutionally staged cityscape (infrastructure, buildings and landmarks) but via the socially scripted meaning attributed to space. While social actors can rescript space in a way that traverses divisions, social scripting may also correlate with institutional divisions. As previously noted, the effect of social movement or individual perception of space varies from person to person, while shared experiences form social knowledge. The variability of meaning ascribed to places is considered by Lefebvre, who envisages existence not;

> as an architecturally rigid edifice, or as a flowing river, but as a colossal interaction of levels, from the subatomic to the galactic...from small

social groups to the large sociocultural formations we call "civilisations". (Lefebvre 1991: 120)

While Lefebvre (1991: 120) regards these conceptual and physical spaces as 'levels', Mitchell refers to the 'the wide set of human activities that involve the creation, preservation and alteration of multiple worlds' as the process of 'world-building' (2011: 3–4). However, there is a distinct variability to this process which is not encompassed by either 'levels' or 'world-building' as there is a fluidity in the transformation of space (Lefebvre 1991: 120; Mitchell 2011: 3–4). Though levels may be distinguished in the staging of space, as institutional actors direct the use of city space, there is an interactivity between the institutional and the social, through top-down urban planning constructs, and the social bottom-up utilisation of city space. The framing of institutional and capital, as actors and tools of staging and restaging, conceptually acknowledges the influence of these actors in physically altering spatialities. While the capability of social and transgenerational actors to engage in scripting and rescripting narratives and meaning in space demonstrates their agentive capabilities. As noted previously, agency is strengthened socially, which demonstrates its spatial dynamic. The spatialisation of agency is important to observe with regard to poltical divisions in Mostar, while experiences of space are momentary, the greater number of social actors using a space in a particular function strengthens and reaffirms the scripting of the space.

Rescripting and Restaging Space

The 'equation' presented in Chapter 1 (Fig. 1.1) outlined the relationships between the processes of rescripting and restaging and the actors who perform the processes. These processes take place, and are performed, within the four spaces of peace, personal, social, urban, and institutional. What follows will illustrate the interactivity of the conceptualisations of rescripting and restaging within the spaces of peace. To recap, social and transgenerational actors facilitate rescripting of space, while institutional actors with capital are able to restage space. Spaces are scripted multiple times and are edited collaboratively and constantly. This is similar to understanding peace as a process and not simply a "place", accordingly, spaces of peace are not strictly geographical or conceptual, instead they are liminal, and this relationship is mutually sustaining.

While the staging of space can be in one function, the scripting can establish a different function. Therefore conceptual and geographical narratives of space may not align, and multiple social narratives exist in space.

As Fig. 1.1 presents, institutionally, cities are staged by top-down powers. The inclusion of capital in the equation for restaging is due to the inherent importance of monetary dimensions to the establishment of space. While transgenerational actors, are pertinent in the transformation of space as their use and movement of space can confirm the scripting across time. Transgenerational actors are of course, social but are included as distinct to acknowledge the way in which memory and movement can maintain spaces across time. Fundamentally, social scripting often does not typically change the physical structure of space, but it generates a narrative of the area reflecting the ability for social actors to transform staged space in subtle ways. The problem becomes a type of chicken-egg scenario, what came first, structure or agency? Ultimately, it is hard to disentangle the two, as Giddens (1979: 69) observes through the theory of structuration there is a mutuality, both structure and agency support and maintain each other, even if they transform each other in the process. This very much reflects social use of place, indeed, place itself is one of structure and typically a home to structural powers whereas space is fluid. Unpacking the spatial context of rescripting and restaging, social actors theoretically have the greatest power of influence due to the legitimacy social actors provide to the institutional and monetary actors. In turn, institutional actors are influential in their capability to restage spatialities. Social actors transgenerationally legitimise power structures, though the consensual or agentive nature of this is debateable in all contexts. However, there is a reason why protests are frequently squashed by institutional powers. As Margaret Mead cautions 'never doubt that a small group of thoughtful, committed citizens can change the world' (in Bowman-Kruhm 2003: 141). In focus, the potential of protests (in this example, non-violent ones) to transform the 'power-geometries' (Massey 2005: 166) of space is explored by Chenoweth and Stephan (2012: 6) through a data set of '323 violent and nonviolent resistance campaigns between 1900 and 2006'. Over this time period Chenoweth and Stephan observed an increase in the frequency of nonviolent campaigns and an increase in their successes, and though violent campaigns still occurred their success rate declined (2012: 7). Fundamentally, nonviolent 'anti regime campaigns' had a much greater chance of success (Chenoweth and Stephan 2012: 7). What this demonstrates is the power of the collective in shifting power, and also implicitly, narratives and even the politics of place. Such protest initiatives are usually spatially orientated and use public spaces to block or dissuade movement or use of spaces and to enhance the visibility of their cause. Accordingly, this framework observes the social actor as being the most influential, due to the influence social actors have in legitimising space. As Lefebvre (2009: 26) proposed '([s]ocial) space is a (social) product' and as such, it is 'constructed' by social interactions. These take place in space and are transgenerationally scripted and staged by institutional actors and capital. Social actors in the context of this work are the residents of Mostar, without institutional or monetary abilities to physically transform space (through renovation or reconstruction). As previously noted, the city space is regarded by Björkdahl (2015: 117) as divided via 'spatial governmentality' which operates as an interwoven and complex process;

> where antagonistic ethno-nationalist identities are spatially expressed and space is contested along ethno-nationalistic cleavages, spatial governmentality techniques re-inforce these divides in order to produce ethnoscapes.

In Mostar, the spatialization of divides in the city reflects the top-down processes of the Dayton Agreement in 1995, and the maintenance of peace through the city statute in 2004. Urban development can be compared to liberal peace processes through the creation of spaces of engagement, which are directed top-down. Liberal peace processes aim to create peace spaces for shared usage, but in practice, this may generate more conflict. Nevertheless, institutional actors are the core instigators of restaging a specific spatiality. While social and transgenerational actors script and rescript meaning in place through usage. The potential for individuals to transform the use of city space can be further discussed from Kappler's (2014: 24) analysis of meaning attributed to space, in that such meanings;

> in turn, represent the structural context of action on one hand, in terms of impacting upon how actors perceive and make sense of their environment, while, on the other hand they also point to actors' ability to challenge this structural context by creating new meanings.

Fundamentally, there is a material dimension to the conceptualisations of structure and agency. Social performances in space can depart from the narrative of the stage through the script social actors generate via movement. As discussed, all of the categories are social in nature, but the word "social" is used primarily to refer to the residents of the city. Through movement residents of divided cities can reaffirm or traverse divisions. Therefore, social movement has the potential to traverse the institutional staging and establish social relations which transcend the conflict divisions. Significantly, all space can be divisive in some respect as we establish boundaries in our everyday lives through movement. Accordingly, it is important to reflect on the complex narrative of even seemingly open space through boundaries inherent in space, and to investigate the staging and scripting of these sometimes-invisible divisions, in order to understand how rescripting experiences of space may transform these margins of movement. While social actors may direct their own movement through spatialities, they typically do not construct the stage and urban spaces are staged in their function through centres of consumerism and the structured cityscape. Additionally, the use of spaces by social actors can be understood through different sub-divisions (for example class, gender, ethnicity) and spaces become scripted in their function via social and transgenerational performances. To summarise these variables can enhance or limit the capabilities of social actors to engage in city space.

In reflecting on the importance of narratives and perceptions of space, (besides that which the individual finds themselves) all space exists as a space of presumed action. Therefore, while experience can be indicative of the future experience of place, it can never be certain as experiences of space are time and context dependent. As such, all space can be considered to be to a certain extent, what Said (1979: 49) conceptualised as an 'imaginative geography' whereby distant assumptions or conceptions regarding actuality of 'place' are constructed. This is a process detectable at a local, national, and international level in the production of narratives; but any space not in the immediacy of the individual self, exists as largely 'imagined' in some respect (Said 1979). As such, spaces are maintained by the emplacement of social expectations, which may be culturally or spatially informed via previous social movement. It is through the imagined narratives of the defined purpose of space, that the potential for directed narratives regarding performed ownership of physical space can be observed. Therefore, the personal narratives within institutional, urban, social spaces, can enhance understanding of how social actors interact with the staged city space. And therefore, how social movement may enact conflict transformation through the rescripting of space.

Spatial transformation, which may be reconstruction or destruction, can occur within and between a multiplicity of spaces. For example, the 1992–1995 intrastate conflict transformed institutional, urban, and social spaces. The conflict played out as part of the dissolution of the former place of Yugoslavia; violent conflict emerged in numerous newly defined places including BiH, in the area generally referred to as the Balkans. Critically, the conflict was initially framed by US president Bill Clinton as a 'European issue' (Horvitz 13 May 1993). This categorisation placed accountability of the emerging crisis as one to be solved by European powers exclusively, and 'We Don't Do Mountains' was circulated as 'summarising the U.S Army's initial resistance to sending troops to Bosnia' and later to Kosovo (Kaplan 2012: 17). However, it is notable that BiH was not, and is not, located within the European Union. Therefore, the act of emplacing BiH, as a responsibility of Europe, though notably not part of the EU, illustrates Said's concept of 'imagined geographies' in the institutional, and conceptual-physical positioning of BiH (Said 1979: 49). Dually so, in BiH, at a time of limited communication methods, the decisive cutting of telecommunication and infrastructure services during the 1992–1995 war created a vacuum of information. This void was filled by institutional narratives which instrumentally constructed the threat of an imagined 'other' (Said 1979). The role of elite manipulation and instrumentalism in the mobilisation of individuals during the war is important to consider in order to trace the interactivity of the spaces of peace and actors of rescripting and restaging. The top-down power dynamic inherent in the instrumentalisation of the conflict is a key aspect in noting the significance of institutional actors in the restaging of spatialities. Consequently, institutional actors can be distinguished from social actors due to distinct power capabilities in their roles (sometimes referred to as elites) and although these individuals are social by nature, the motivations for their actions are congruent with institutional roles and their maintenance of power. It is also notable that through conflicting narratives of the ownership, heritage and use of the physical space, elite conceptualisations of space restaged the map of the Balkans which shaped the conflict and the post-conflict space.

INSTITUTIONAL, URBAN, AND SOCIAL PEACE

The positionality of the institutional is important to understand regarding the influence of elite manipulation in escalating conflict through narratives of place identity. In particular, in the instigation of the 1992–1995 conflict in Bosnia, contested concepts of the heritage of place can be identified. It is important to note that the set pieces of historical narratives of place heritage, such as 'flags, images, ceremonies and music', which are used to invoke the conception of citizenship are 'largely invented' (Hobsbawm 2012: 12). The directed and staged nature of this is evident in Hobsbawm's conceptualisation of 'inventing traditions' which involves 'a process of formalisation and ritualisation, characterised by reference to the past, if only by imposing repetition' (Hobsbawm 2012: 4). This process utilises 'history as a legitimator of action and cement of group cohesion' and in the context of a divided city (and country), historical narratives through education and politics seek to transgenerationally influence social movement, which reaffirms political narratives of heritage (Hobsbawm 2012: 12). Here the link between Massey's (1994: 265; 1991: 27) spatial moments, meeting and weaving can be connected to the establishment of collective memory (Halbwachs 1992). The process whereby individuals anchor themselves to a particular space is notably influenced by 'political myths' fostered (or manipulated) through collective memory which in the context of BiH have been reaffirmed transgenerationally (Bottici and Challand 2012: 24; Halbwachs 1992). Notably, the utilisation of 'customs, traditions, and language in order to promote and reinforce a particular ethnic community' is detectable in the ethno-nationalistic political narratives of the president of the Socialist Republic of Serbia, Slobodan Milošević, and reciprocal actions by Croat leaders' including Franjo Tuđjman, the president of Croatia (Jesse 2014: 95). Critically, the creation of 'political myths' are a core tool in establishing control through instrumentally inflaming tensions for the benefit of the political elite (or institutional actors) (Bottici and Challand 2012: 24). Such myths function as 'a drama on stage' and are influential through an ability to 'coagulate and reproduce significance' to social actors (Bottici and Challand 2012: 24). The mobilisation and militarisation of citizens during intra-state conflicts and the persistence of conflict narratives transgenerationally, evidences the capability of institutional actors to direct social roles and movement. In this capacity, the longevity of such narratives exist as 'mapping devices' which spatially inform how individuals 'feel about [the world] and also act within it as a social group' (Bottici and Challand 2012: 24). The persistence of such political myths (Bottici and Challand 2012: 24) in BiH, can be observed through the Dayton Agreement and the establishment and maintenance of divided education in the country. The divided education system not only presents students with specific historical narratives, but the segregation also directs social movement and use of space. Therefore, institutional actors play a functional role in not only de-escalating conflict but also restaging the spatiality.

In BiH, the signing of the Dayton Agreement in 1995 ended the active military conflict which had transformed the spatiality. The signing of the accords facilitated a cessation of the ethno-nationalistic violence which had started in 1992. This institutional agreement meant that militarised and mobilised groups which operated ceased military action. The top-down process of defining the spatiality as "in peace", in turn impacted on the urban demographics and social movement in the spatiality. What is notable in contemporary BiH, through the establishment of the Dayton Agreement, is that the Constitution involves the staging of identity. This is observed by Markowitz (2010: 78), as the 'the Bosnian state', directs citizens 'to think and act in terms of a tripartite citizenry comprised of members of one and only one of the Bosniak, Croat or Serb constituent nations'. This illustrates the previously noted importance of institutional peace in the cessation of the Bosnian war, due to the instrumentalised divide of the conflict. And also demonstrates ongoing political narratives which attempt to stage the country, as socially divided. Though the circumstance of institutional peace has a direct impact on social and urban spaces, the compliance or agreement of social actors in the urban (and non-urban) spaces demonstrates that social and the urban peace spaces complete institutional peace.

Notably an 'urban' environment can be considered one that involves a constant state of conflict transformation wherein social actors from different backgrounds and cultures meet on a stage defined by top-down planning. This interaction may be a continuous and conscientious active 'conflict', but notably one without 'violence' (Mitchell 2011: 7). This state of peace, or conflict without violence, is somewhat assisted via institutional means (in the 'urban' setting) with the availability of infrastructure and sufficient resources (Mitchell 2011: 7). Spatially, the urban represents a series of unique yet interconnected and interrelated spaces which facilitate movement. There is an instrumentalised cohesion in such disparate spaces as observed by Said (1979: 49) through 'imaginative geographies' and Anderson (2006: 6) through 'imagined communities'. Imagined geographies of communities were identifiable in the instigation and militarisation of the 1992–1995 Bosnian war. Crucially, these spatialised narratives are performed in, and were for the purpose of securing (institutionally staged), urban spaces. Fundamentally, urban space is socially dense and numerous sub-divisions are observable. For example, capital is an important variable for social actors to access and engage with some spaces. While this is of importance contemporarily in the city, it also directly impacted on experiences during the 1992–1995 war, as it was common for individuals to leave the physical space of the conflict, if they could afford to do so (Mertus et al. 1997). Through this, the variable of economic class and the institutional directing of conflict to 'divide the working class'; and to generate conflict turning 'workers against themselves' can be observed (Jesse 2014: 95). This perspective, similar to other constructivist theories, highlights other variables in the conflict, looking beyond the exploited, staged, 'ethnic differences' (Jesse 2014: 95). Furthermore, this highlights the impact of 'group inequality, discrimination in jobs, [and] education' and the propensity for 'one group' to politically 'dominate' the other which has the potential to exacerbate divisions (Jesse 2014: 95). These remain important issues in discussing social movement in the post-conflict city.

From neo-liberal peacebuilding perspectives, peace is obtainable through economic means (Paris 2010: 341). However, this significantly overlooks the readily accepted competition inherent in the capitalist system, and through divisions of labour in which conflict (via competitive consumption) is a functional tool for generating economic growth (Paris 2010: 341). To lend a term from Galtung, capitalism itself instigates a negative peace, in many ways this can also be structurally violent and is frequently materialised in physical violence (Galtung 1967: 12; Galtung 1990: 291). Fundamentally, as discussed previously, money or capital is integral in the physical construction of 'place', space equals power and construction in space is funded by capital.

In post-conflict environments and some spaces are prioritised over others in reconstruction. Fundamentally, the social anchoring of the self to physical place, is reinforced through the (forced) consumption of 'political myths' which produce spatial belonging, this narrative of belonging and heritage of land is then sometimes used to instigate conflict (Bottici and Challand 2012: 24). The spaces which establish these narratives are typically those that are reconstructed first post-conflict, which does not always benefit social actors. There is a complex relationship between conceptual and physical spaces of peace and the actors using, creating, and maintaining such spaces which represents an important dimension in exploring the use of post-conflict social space. Notably, the divisions in the city of Mostar impacts on the everyday lived reality of residents through the social divisions in cultural and sporting groups for Croat and Bosniak citizens, as well as a divided education system. Invariably, everyday 'conflict' takes place in all cities, but this can be exacerbated by intrastate conflict. While the conditions for social peace are in constant flux, the staging of space as shared can be better established by institutional actors in post-conflict divided spaces through consultation with social actors. Despite the trickle-down divisions enshrined by the Dayton Agreement social actors do establish shared spaces in the city of Mostar by creating narratives of space that do not correlate with institutional definitions of space. As discussed, the three overlapping spaces of peace; the institutional, urban, and social (with the personal space collapsed into the social) can interact but often do not. By drawing these spaces together and unpacking the movement within, the interactivity of the two processes of: restaging and rescripting can be illustrated. The interactivity of physical and conceptual spaces of peace is explored further in chapter three.

NOTE

1. Goffman uses a gendered example of a surgery going wrong 'when a surgeon and his nurse both turn from the operating-table and the anaesthetised patient accidentally rolls of the table to his death, not only is the operation disrupted in an embarrassing way, but the reputation of the doctor, as a doctor and as a man, and also the reputation of the hospital *may* be weakened' (Goffman 1971: 235, emphasis added).

BIBLIOGRAPHY

Anderson, B. (2006). *Imagined Communities: Reflections on the Origin and Spread of Nationalism*. London and New York: Verso.

Barthes, R. (1997). Seminology and the Urban. In N. Leach (Ed.), *Rethinking Architecture: A Reader in Cultural Theory*. London and New York: Routledge.

Björkdahl, A. (2015). 'Two Schools Under One Roof' Unification in the Divided City of Mostar. In A. Björkdahl & L. Strömbom (Eds.), *Divided Cities, Governing Diversity.* Lund: Nordic Academic Press.

Björkdahl, A., & Kappler, S. (2017). *Peacebuilding and Spatial Transformation: Peace, Space and Place.* Oxon and New York: Routledge.

Björkdahl, A., & Selimovic, J. M. (2015). A Tale of Three Bridges: Agency and Agonism in Peace Building. *Third World Quarterly.* Available from: http://www.tandfonline.com/doi/full/10.1080/01436597.2015.110825. Accessed 12 Feb 2016.

Bottici, C., & Challand, B. (2012). *The Myth of the Clash of Civilisations.* London and New York: Routledge.

Bowman-Kruhm, M. (2003). *Margaret Mead: A Biography.* Westport, CT: Greenwood.

Chenoweth, E., & Stephan, M. J. (2012). *Why Civil Resistance Works: The Strategic Logic of Nonviolent Conflict.* New York: Columbia University Press.

Foucault, M. (1980). Questions on Geography. In C. Gordon (Ed.), *Power/Knowledge: Selected Interviews and Other Writings, 1972–1977.* London: Harvester Wheatsheaf.

Foucault, M. (1991). *The Foucault Effect. Studies in Governmentality. With Two Lectures and an Interview with Michel Foucault* (G. Burchell, C. Gordon, & P. Miller, Eds.). Chicago, IL: University of Chicago Press.

Galtung, J. (1967, September). *Theories of Peace: A Synthetic Approach to Peace Thinking.* Oslo: International Peace Research Institute. Available from: https://www.transcend.org/files/Galtung_Book_unpub_Theories_of_Peace_-_A_Synthetic_Approach_to_Peace_Thinking_1967.pdf. Accessed 25 Nov 2014.

Galtung, J. (1990, August). Cultural Violence. *Journal of Peace Research, 27*(3). Available from: http://www.jstor.org/stable/423472. Accessed June 2016.

Goffman, E. (1971). *The Presentation of Self in Everyday Life.* Harmondsworth: Penguin.

Giddens, A. (1979). *Central Problems in Social Theory: Action Structure and Contradiction in Social Analysis.* Berkeley and Los Angeles: University of California Press.

Gupta, A., & Ferguson, J. (1992, February). Beyond "Culture": Space, Identity and the Politics of Difference. *Cultural Anthropology, 7*(1), 6–23.

Halbwachs, M. (1992). *On Collective Memory.* Edited, translated and with an introduction by L. A Coser. Chicago, IL: University of Chicago Press.

Hobsbawm, E. (2012). Introduction: Inventing Traditions. In E. Hobsbawm & T. Ranger (Eds.), *The Invention of Tradition* (pp. 1, 4, 12). Cambridge: Cambridge University Press.

Horvitz, P. F. (1993). Intervention in Bosnia: Clinton Mutes His Fervour. *New York Times,* Archives. Available from: http://www.nytimes.com/1993/05/13/news/13iht-poli_7.html. Accessed 20 Nov 2014.

Jesse, N. G. (2014). Ethnicity and Identity Conflict. In K. De Rouen & E. Newman (Eds.), *Routledge Handbook of Civil Wars*. Oxon: Routledge.

Kaplan, R. D. (2012). *The Revenge of Geography*. New York: Random House.

Kappler, S. (2014). *Local Agency and Peacebuilding: EU and International Engagement in Bosnia-Herzegovina, Cyprus and South Africa. Re-Thinking Peace and Conflict Studies*. Basingstoke: Palgrave Macmillan.

Lefebvre, H. (1991). *Critique of Everyday Life* (Vol. 2). London: Verso.

Lefebvre, H. (2009). *The Production of Space* (D. Nicholson-Smith, Trans.). Oxford: Blackwell.

Markowitz, F. (2010). *Sarajevo: A Bosnian Kaleidoscope*. Chicago: University of Illinois Press.

Massey, D. (1991, June). A Global Sense of Place. *Marxism Today*. Available from: http://banmarchive.org.uk/collections/mt/pdf/91_06_24.pdf. Accessed 20 June 2016.

Massey, D. (1994). *Space, Place and Gender*. Cambridge: Wiley.

Massey, D. (2005). *For space*. London: Sage.

Massey, D. (2007). *World City*. Cambridge and Malden: Polity Press.

Mertus, J., Tesanovic, J., Metikos, H., & Boric, R. (1997). *The Suitcase: Refugee Voices form Bosnia and Croatia*. Berkeley: University of California Press.

Mitchell, A. (2011). *Lost in Transformation: Violent Peace and Peaceful Conflict in Northern Ireland. Re-Thinking Peace and Conflict Studies*. Basingstoke: Palgrave Macmillan.

Newman, D. M. (2013). *Sociology: Exploring the Architecture of Everyday Life*. Los Angeles: Sage.

Paris, R. (2010). Saving Liberal Peacebuilding. *Review of International Studies*. Available from: http://www.engagingconflict.it/ec/wp-content/uploads/2012/06/Paris-Saving-Liberal-Peacebuilding.pdf. Accessed 20 Dec 2013.

Said, E. W. (1979). *Orientalism*. New York: Random House.

Simmel, G. (1997). *Simmel on Culture: Selected Writings* (D. Frisby & M. Featherstone, Eds.). London: Sage.

Tonkiss, F. (2005). *Space, the City and Social Theory: Social Relations and Urban Forms*. Cambridge: Polity Press.

CHAPTER 3

Spatialising Conflict Transformation: *Spaces of Peace (and Conflict)*

SPATIAL AND CONCEPTUAL [CONFLICT] TRANSFORMATION

Conflict transformation is to envision and respond to the ebb and flow of social conflict as life-giving opportunities for creating constructive change processes that reduce violence, increase justice in direct interaction and social structures and respond to real-life problems in human relationships (Lederach 2003: 22).

Space is the vanguard of human settlement, it can provide visibility, security and it also exists as the stage for a narrative of heritage to be scripted. Through Said's concept of 'imaginative geographies' and Anderson's conception of 'imagined communities' the conceptualisation of physical space as states, can be seen to be socially constructed or 'imagined' (1979: 49; 2006). It is notable that all conceptions of space, from the local to the international, are in some respects imagined. The importance of ownership, heritage and use of 'place' is therefore a central issue in identifying the occurrence of intra-state conflicts. For example, the division of the formerly recognised spatiality of Yugoslavia led to the re-definition of the physical space of BiH. Fundamentally, political manoeuvring led to the breakup of the previously conceptualised space of Yugoslavia which instigated the 1992–1995 conflict; and the fact that BiH was, and is, not within the political parameters of Europe, affected the duration and international responses to the conflict. Though briefly outlined here, the impact of the restaging of the space had far reaching

© The Author(s) 2019
S. Forde, *Movement as Conflict Transformation*, Rethinking Peace and Conflict Studies, https://doi.org/10.1007/978-3-319-92660-5_3

implications which demonstrate that the link between space and security can be observed as not simply one of topography, but one of international relations.

IDENTITY AND PLACE

Space becomes place through experience and social memory, place is therefore both an intrinsically individual and yet collective experience. Places may evoke 'topophilia or topophobia' (the affection or aversion felt for a place respectively) (Tuan 1974). This occurs in space and through social movement, according to Lippard (1997: 10), 'if space is where culture is lived, then place is a result of their unison'. Critically, 'places are places...because they have identity' (Hague 2005: 7). Therefore, spaces become places through social movement, and through movement social actors rescript the future use of the space. This can be examined through sociological, political, and geographical paradigms that, when considered together, construct a comprehensive understanding of the not only the restaging but also the rescripting of place.

In everyday life, we move within and between spaces generating our own personal borders of movement and contributing a social script to institutionally staged spaces. However, such borders are not typically instrumental. More formally, passports allow for movement between institutionalised borders, this bureaucratisation of identity is one that is linked to geographical space upon which a matrix of identity is formed. Narratives of ownership of space dominate political discourse and a global increase in political narratives which seek to problematise movement (or certain types of movement) can be considered a reactionary backlash to increased global social movement (Massey 1991). Moreover, the categorisation of identity through nationality can be linked to increasingly global interactions, as institutional documentation verifies and legitimises the individual in a world of relative strangers. This rationalisation of identity is one that is constructed of individual and social expectation. Place is a result of social movement and use of space, however, due to historical institutional restaging and social rescripting, there may be competing narratives of the history of the space of the place. Such narratives of ownership create a narrative border alongside, or in support of, the establishment of a physical border. Largely, these borders are conceptualised through the institutional narratives, but also economically and socially. "Imagined" borders can be identified in the city of Mostar, as there is no physical division in the city to divide Croat and Bosniak city areas, but these are materialised socially through, largely informal, signs and signifiers, and more formally materialised through public services, and the divided education system.

IMAGINED BORDERS

Instead, then, of thinking of places as areas with boundaries around, they can be imagined as articulated movements in networks of social relations and understandings. (Massey 1991: 28)

In this above quote, Massey discusses the way in which place can be understood as a socially fluid experience, spatialised on 'a street, or a region or even a continent', but having a ripple effect far beyond the physical context (Massey 1991: 28). Moreover there is a fluidity to the social construcation of places, as places can be understood to be constellations of social interactions (Massey 1991: 29). The perpetual motion of these interactions can be observed physically and theoretically. While certain interactions form place, these interactions are not static and as such can transgress and transform staged space. Therefore, it is important to recognise the potential of social movement as demonstrating the power of social actors to rescript space. Through this Massey's (1994: 121) objective of looking at social relations and spatial organisation differently can be operationalised. Understanding the rescripting of space importantly, reflects not 'how places *are* currently seen...but how places *could* be seen' (1994: 121) Fundamentally, Massey (1994: 121) puts forward that an 'anti-essentialist construction' understanding of space and identity can highlight issues with 'any automatic associations with nostalgia and timeless stasis' regarding place identity. This essentially unpacks and exposes the empty narratives of the so called 'authentic character of any particular place' (1994: 121). Through this, Massey (1994: 121) presents important points regarding the socially constructed nature of heritage, identifying the power dynamics in the construction of such narratives, and outlining an important misconception of place as timeless and static. Critically, Massey (1994: 122) proposes the symbiotic relationship of the spatial distribution of social interactivity which;

forces us to recognise our interconnectedness and underscores the fact that both personal identity and the identity of those envelopes of space-time in which and between which we live and move...are constructed through that interconnectedness.

Fundamentally, this demonstrates the relevancy of the interlinkages of social interactions, and the construction of place, and the continued narrative of place. Overall, it stresses the blinkered perspective of narratives of identity which overlook the spatial affinity of humanity. As observed in the ethno-nationalistic divisional narratives of the 1992–1995 Bosnian war, politicised narratives of identity traditionally seek to establish functional spatial boundaries. In Mostar, the disputed narratives over the ownership of space led to the division of the city between East and West and resulted in the forced displacement of citizens from the West of Mostar to the East, and from the East to the West (Calame and Charlesworth 2011: 104). Following the cessation of violence, the subsequent post-conflict agreements and city statutes have reaffirmed the institutional narrative of the conflict division. Therefore, due to ongoing ethno-nationalistic governance, the space is staged as divided.

The boundaries of the divided city of Mostar are evidenced institutionally, in education, and through local facilities such as the postal service.[1] As previously noted, in everyday life we create spatial boundaries of movement, these boundaries also define and segregate spatial landmass. It is relevant to note that, in the post-Yugoslav space, the establishment of boundaries was a central issue in the militarisation of the conflict. Therefore, not discounting social capabilities to rescript space, institutional actors can be considered to hold considerable power in restaging space, specifically in the creation and maintenance of borders. Yuval-Davis (2011: 95) reflects on Massey's above discussion of borders, disagreeing on the conceptualisation regarding 'the relative (lack) of [the] importance of borders and boundaries'. However, Yuval-Davis (2011: 95) observes the resonance of Massey's outline of the understanding the fluidity of the narrative of space as 'the same location can be constructed in totally different—and sometimes contradictory nationalist narratives'. Borders, in short, are spatially staged boundaries imbued with 'symbolic resonances' which become part of the 'founding myths of the state'-through this they create narratives of division staging 'the world into "us" and "them"' (Yuval-Davis 2011: 95–96). While political narratives in BiH are divisive and maintain divisions from the conflict for political means, it is notable that borders and boundaries are also a functioning part of everyday life. Such borders may be physical or conceptual, or may be temporal. Fundamentally, there is a social fluidity to the construction of place, as Massey (1991: 29) observes 'places are processes'. As such, critically, there is no definitive ending in the construction of place, signalling the potential of place and space to be rescripted innumerable times by social actors. Places, as observed by Massey (1991: 29) are 'constructed out of a particular constellation of social relations, meeting and weaving together at a particular locus' (Massey 1991: 27). Much like the constellations in our own skies, sometimes we may not be able to see these them (as previously noted in Chapter 1 this links to the time we spend looking), and most likely we will not be around long enough to see many of them change, but they are there nonetheless and their transformation is an inevitable process. Places can be thought of as a multiplicity of spaces, without boundaries; however, this does not mean they are homogenous and as such, are naturally spaces of conflict (Massey 1991: 29). Socially, we are interconnected, however, there are barriers to understanding and contextualising these relations. Our socialisation has been constructed around a division of space, in the home, the front door, outside, inside; socially different spaces hold different functions and our spatial access to them has a great deal to do with positionalities. In short, we are interconnected, left, right, forward, backward, up, and down, our socio-spatial trajectories influence our spatial affinities which can lead to conflict though this may not be physically violent. The question is then how do "everyday" conflicts play out in a divided city space? Such spatial interactions can impact on movement, as Kappler (2014: 36) observes, perceptions and understandings of space impacts on how social actors use space, and therefore impacts on how a space is scripted or can be rescripted.

Conceptually Mapping Conflict

In 'The Little Book of Conflict Transformation', Lederach compares explaining a conflict to a topographical 'relief map of the peaks and valleys' of a situation (2003: 8). In this comparison the peaks represent 'significant challenges in the conflict', and the valleys represent 'failures, [and] the inability to negotiate adequate solutions' to the conflict (2003: 8). Through this metaphor, Lederach notes that, in addressing conflict, there is a tendency to overlook 'underlying causes and forces' which are foundational (Lederach 2003: 8). Lederach (2003: 9–11) also discusses the importance of understanding the complexities of a situation in that 'looking requires lenses that draw attention and help us become aware' whereas 'to see...is to look beyond and deeper'. Further still, through conscious observation and analysis, 'conflict transformation' is a way of both, 'looking as well as seeing', both of which 'require lenses' (Lederach 2003: 9–11). With this in mind, Lederach highlights the importance of understanding 'the *content*, the *context*, and the *structure*' and that through understanding this, it is then possible for conflict to instigate constructive change (Lederach 2003: 12).

As discussed briefly in Chapter 2, there are four overlapping spaces (institutional, urban, social, and personal) in which conflict transformation can be enacted. The complexity of these interacting spaces is demonstrated in Mitchell's proposition that 'human beings occupy multiple worlds' that exist as 'unique ontological spaces in which different people act, experience their lives and interact with others' (2011: 3). In Mitchell's conceptualisation, the influence of social actors (and transgenerational actors) can be observed in the formation of the meaning attributed to the 'world' (Mitchell 2011: 3). However, the importance of the physicality of the spaces is also noted as 'physical places, objects and institutions act as its medium' (Mitchell 2011: 3). Through the framework of this book, it can be surmised that physical space is the stage on which social actors perform the intangible social memory of space-based experiences. Observably, social actors create their own narratives in space and multiple narratives can exist of any one space. The world in which we find ourselves is notably conceptualised via different spaces and places which are staged within the physical topography of the land, but these also have a milieu of social rescriptings.

The legitimisation of physical space as place is directly influenced through the restaging and rescripting provided by institutional and social actors respectively. The conceptual and physical spaces (institutional, urban, social, and personal), are constructed, maintained and legitimised in their function by distinct social performances. Fundamentally, social actors legitimise these conceptual and physical spaces via their social performance towards, and within these spaces. Individuals operate under the institutional, within the urban space, through social and personal spaces, and all movement involves a renegotiation of different identities. For example, while a resident of Mostar lives within the jurisdiction of a Bosniak or Croat city area of Mostar, they may move across the institutional divisions in the city but also may not. In this capacity, socially and personally, individuals use and script a distinct narrative of spaces which, through movement, has the potential to rescript experiences of space and importantly, relations with other social actors in the spaces.

SPACES OF PEACE

This nested diagram (Fig. 3.1) presents the proposed relationship between the spaces of peace. The institutional space is legitimised by the urban, social, and personal spaces (for analysis, the space of the personal is integrated into the social space). Fundamentally, the position of the spaces in the diagram represents their relationship in conceptual, and physical space. In particular, the social space encompasses the other spaces. All spaces are social, and the social space facilitates the legitimisation and creation of the other spaces. The capabilities of social actors are evident in the space of the urban, as it is a space that is populated, scripted, and used by social actors. Similarly, institutional spaces, such as governmental buildings and jurisdictions, exist due to the legitimisation by social actors. Social actors, through the use of space, confirm or rescript the staging of the space. For example, following the 2014 protests across BiH damage to the governmental buildings prevented the use of the buildings in their staged functioning (ICG 2014: 4). While social actors, generally even those in government, cannot individually

Fig. 3.1 Nested diagram of conceptual and 'physical' spaces of peace

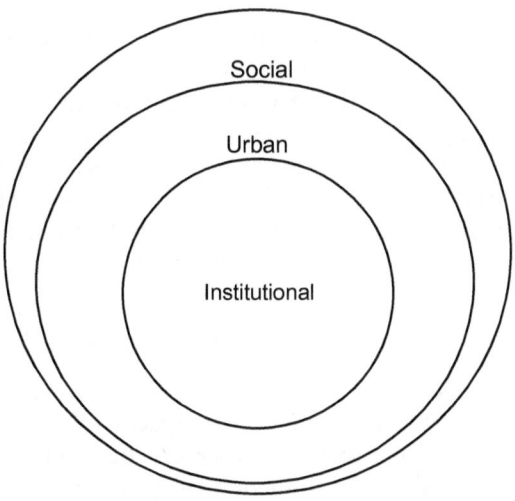

choose to change the official staging of spaces in city, they can collaboratively transform the use of the space. Notably, the urban space is the space wherein institutional planning and social movement interact. It then follows that the social and personal spaces of peace influence the use of the urban space. Though separated into spheres, these mapped out spaces of peace are interconnected, physically and conceptually.

When a conflict ends, it is typically in the form of a 'negative peace' through the cessation of violence and ideally a transition to a 'positive peace' occurs (Galtung 1967: 12). As structures of governance are socially legitimised, a top-down agreement of the cessation of a conflict, through formalised peace agreements, begins the process of peace. This then stages the wider spatiality in peace. This staging ends physical violence and changes the staging of the space as transitioning to a time defined as "post-conflict". Reflecting on structure and agency, it can be noted that institutional structures stage space, however, social agency has the capability to direct the narrative and use of staged space. In this capacity, it is worth noting that the capability of individuals to exercise agency can be curtailed by structure. In focus, individuals can be manipulated by institutional actors, for the creation and maintenance of a conflict. As such, the institutional space of peace serves a core function in the establishment of peace. The 'positive peace' which ideally follows, is conceptualised by Galtung (1967: 12) as 'a synonym for all other good things in the world community, particularly cooperation and integration between human groups'. In the realisation of this integration, the location and importance of access to resources is fundamental in deeply divided societies (Galtung 1967: 12). According to Galtung's (1967: 12) theorisation, the Dayton Agreement can be analysed as only functioning enough to provide the conditions of a 'negative peace'. Theoretically, the physical spaces of urban and social peace, have the potential to provide a lived 'positive peace' (Galtung 1967: 12). However, these spaces do not exist in isolation from divisive political structures, herein the challenge social actors face in rescripting space is evident.

INSTITUTIONAL PEACE

As discussed, post-conflict spaces are restaged from sites of active violence when in conflict, to spaces of peace following the conflict. An institutional space of peace can be staged on paper, this has previously involved the introduction of liberal principles which aim to instil peace through democracy, and economic involvement. In BiH, the Dayton Agreement brought an end to the 1992–1995 war, and the country was internationally-institutionally defined as a space of peace. The agreement was the first step in the 'liberal state-building process' which, theoretically, aimed to 'heal ethnic differences and eventually recreate a multi-ethnic state through democracy, human rights and a liberal economic system' (Richmond and Franks 2009: 58). This form of peace is observed by Paris (2010: 341) as being implemented on the central assumption 'that rapid [institutional] liberalisation would create conditions for stable and lasting peace in countries emerging from civil conflict'. This conceptualisation of peace was considered to be the complete package of the complimentary institutional processes of 'democratisation and marketisation', and that once the precursors of liberalisation had been set in motion 'they would be largely self-perpetuating' (Paris 2010: 341). However, the internationally led institutional agreement constitutionally, and spatially staged BiH and Mostar as ethnically divided, and entrenched the conflict divisions. Institutional peace provided through liberal state-building or peacebuilding in isolation can be observed to be a peace that is inherently 'negative' as the cessation of conflict does not always entail a transformation of conflict cleavages. As Galtung (1976: 297–298) observes;

> peace has a structure different from, perhaps over and above, peacekeeping and ad hoc peacemaking...more specifically, structures must be found that remove causes of wars and offer alternatives to war in situations where wars might occur.

Though liberal peace aims to connect with the local, in application, it remains disconnected. The top-down liberal peace is legitimised institutionally and internationally, and it is purported to create an institutionally directed sequenced connectivity of the different spaces of peace. This interactivity is essential for conflict transformation to be socially realised but liberal processes frequently fall short. While a top-down institutional agreement stages 'peace', there is often more to be done, as observable in the Dayton Agreement which constitutionally re-enforced, and continues to perpetuate conflict cleavages.

The liberal peace project of the Dayton Agreement, was rationally efficient in pacifying political elites through establishing physical and conceptual spaces of governance in the Republika Srpska and the Federation of Herzegovina. Through this, it also reaffirmed the intractability of the divisions created through the conflict. As Keil and Perry (2016: 5) note, 'post-Dayton BiH' is a spatially fractured, 'ethnically-gerrymandered construction', of which, the top-down legitimised conflict boundaries correspond with 'the dominant political preferences and results of ethnic cleansing'. The Dayton Agreement in BiH notably established a glass ceiling for conflict transformation at an institutional level. While according to Richmond (2012: 29) liberal peacebuilding aims to provide a 'trickle down' peace, the BiH Constitutional divisions present more of a trickle-down conflict through perpetuating the formation of identity via the recognition of the three main ethnic groupings. The peace agreement has transitioned the conflict from actively violent, to a stagnating and structurally violent division (Galtung 1990: 291). This division is one that is reinforced via international and national institutions. It is also perpetuated, socially and transgenerationally in spaces of interaction through predominately ethnically divided residential areas and the divided education system.

It is observable that the the institutional Dayton agreement constitutionally reaffirmed the ethno-nationalistic divisions of the conflict. To this end the deployment of liberal peace can be observed as an 'image of the West...projected where it cannot work' (Chandler 2010: 16). Furthermore, it is a model which is based on a conceptual 'negative peace' of the capitalist system, as one which is based upon accumulation and economic competition (Galtung 1967: 12). This is then transplanted into already fractured societies. The notable failures of the 'one-size-fits-all' paradigm has made the issue of conflict relapse apparent, with implemented structures failing in part in, BiH, Sri Lanka, Sierra Leone, Cambodia, El Salvador, Nicaragua, Rwanda, Liberia, Guatemala, Mozambique, Namibia and Angola (Lewis 2010: 647; Fanthorpe 2005: 27; Richmond and Franks 2009: 65; Chandler 2010: 8). Such failures vested 'too much faith in the liberal subject' (Chandler 2010: 8) and in doing so, fundamentally overlooked other spaces of peace. Critically, 'peacebuilding' in itself 'can be [regarded] an act of violence' while conflict can be a tool for 'resisting, constraining and preventing' violence (Mitchell 2011: 1). Moreover, as peacebuilding occurs in semi-institutional settings, these physical spaces can be seen as a negative extension of external intervention, which has the potential to overlook social needs (Hemmer 1997: 4; Kappler 2013: 356). In particular, institutional peace does not translate to a 'positive peace' when individuals are forced to negotiate displacement, due to the constitutional exclusion of minorities in BiH which renders immediate social needs unsecured (Galtung 1967: 12). The failures of the liberal peace project ultimately led to a refocusing of the post-conflict lens towards localised approaches to peace, and the prioritisation of bottom-up peacebuilding, through the concept of conflict transformation which situates conflict as cyclical and part of being human (Lederach 1995; Galtung 1967: 12). The importance of local involvement in peace and the concept of bottom-up peacebuilding has grown to prominence due to shortcomings of top-down institutionalised 'trickle down peace' (Richmond 2012: 29). As Richmond (2012: 27) observes, there is a recent turn in literature and practice towards 'creating a self-sustaining, civil peace in post-violence, post-conflict environments' which refocuses analysis towards 'the local level'. However, it is telling that the complexities of local contexts are, as Richmond (2012: 27) highlights, often 'relegated to lower (or localised) priorities'. Despite the civil society perception of the 'the local' as a space of agency, the institutionally perceived capability of enacting change through this 'political space' is not reflected in institutional attitudes (Richmond 2012: 27). As Richmond (2012: 27) observes; 'civil society from the perspective of most international actors, represents a method of privatising the service provision often associated with the state'. The result is an extraction of 'local needs and welfare' from the 'state building process' and the marginalisation of culture (Richmond 2012: 27). Fundamentally, this demonstrates that institutional actors perceive the capability of social actors as marginal, but utilitarian in reducing costs and individualising responsibilities. In focus, while local actors are not without agency in post-conflict civil society, it is important to highlight that for institutional actors, this space of agency is a functional displacement of state responsibilities, essentially a form of governmentality (Foucault 1991: 102). Furthermore, as Belloni (2001: 170) observes those who are employed to facilitate the institutional peace project as 'international civil servants' are typically not properly trained and are therefore;

> ill-equipped to comprehend the political, social and cultural contexts within which power relations are expressed when society becomes uncivil and citizens are disempowered and excluded.

To be able to enact 'socio-political change', Belloni (2001: 170) states that 'local groups and organisations have to confront a political struggle in civil society' through engagement with top-down institutions. Recent scholarship cites a growing interaction with these typically international spaces. In particular, Kappler (2014: 171) considers a case study involving the changing conditionality of 'police reform' headed by the 'European Union Police Mission in Bosnia', which aimed to reform policing, and 'centralise the political structures of the country through a multi-ethnic police force'. Following objections 'from a diversity of actors in BiH' some EUPM actors proposed 'the removal of conditionality politics' highlighting the EU's inability to address 'problems related to nationalism' (Kappler 2014: 171–172). Through this Kappler (2014: 172) demonstrates the local agency capabilities of individuals, outlining that as a result the EU 'shifted its approach towards a lighter version of reform as a response to the subtle influence of local actors'. This example represents the sometimes marginal ways in which local agency operates and, also highlights that social actors can influence institutional decisions. To enact against institutional divisional narratives trickling down to a social level Björkdahl (2015: 117) discusses the social importance of the fostering 'shared space' as something that is open to all and importantly exists without an ethno-nationalistic narrative. As Björkdahl and Kappler (2017: 121) observe, the creation of shared spaces demonstrates conflict transformation and among other things a better everyday life for individuals living in post-conflict space. While the urban stage in Mostar is set by institutional actors as divided, it is in this space that social actors may rescript institutionally defined spaces.

URBAN SPACE

An urban place is socially concentrated space, wherein individuals will interact, live and move in close proximity to one another. Bottom-up peacebuilding activities typically occur within urban spaces due to infrastructure and the accessibility of these locations. While this occurs locally, semi-institutional structures form this network of engagement. As a result, top-down international and national power structures, typically direct participation, through establishing the location (and staging) of the shared spaces. Such spaces can facilitate communication, with a potential for a 'transformation' of conflict roles; and though institutionally staged, can represent an opportunity for social rescripting in post-conflict environments (Lederach 1995: 7). As discussed, issues with liberal peace as a top-down directed process have led to a refocusing of research and practice to local levels. As classifications of peace are frequently focused at the institutional level, localised 'conflict transformation' and social rescripting have been traditionally overlooked (Lederach 1995: 7). Bottom-up peacebuilding or 'conflict transformation' therefore focuses on ensuring peace is locally owned; and to ensure that individuals to have a say in what happens following a conflict they are, ideally, given space to 'engage and shape the structures that order their community life' (Lederach 2003: 22). Fundamentally, this underscores the importance of top-down actors working collaboratively with social actors, who are not involved in government or political processes. Such an interaction can facilitate not only spaces of social transformation, but also participatory development or locally led planning of the cityscape (McCarthy 2012: vii; MacDonald 2011: 35). However, the impact of socially led peacebuilding is difficult to quantify and the impact of '"everyday conflict resolution" or everyday diplomacy is often overlooked in the academic and policy literature' (Mac Ginty 2013: 387). Instead, there is a focus on institutionally verified '"experts" and "professionals"', typically external to the conflict environment (Mac Ginty 2013: 387). This directed quantification can overlook the potential of social movement in rescripting and transforming post-conflict staged space, and also social relations within space. As the city of Mostar remains institutionally divided, an urban peace may be then considered through spatial-temporal moments of social rescripting and sub-divisions of movement in the city. This is not to characterise the situation in Mostar as one that is irreconcilably divided but to note that, while sub-divisions may exacerbate the institutional narrative of the ethno-nationalistic division, they may also transcend it.

An urban peace, can be fostered through shared space which facilitates movement across conflict divisions. As previously noted, places are culturally developed spaces, distinct in their historical imprints and social lineage. Urban spaces are institutionally staged and the social lineage of urban spaces is scripted by social movement, and 'narratives of personal memory' (Neisser and Fivush 1994: 90). Transgenerationally the scripting of space can legitimise the staging of space or may result in rescripting. For example, a space is staged as a park, socially it is used as a park, over time or during a conflict the park is damaged and left in a dangerous condition and is not used as a park. It is in the post-conflict environment that ideally the space is restaged as a park, and rescripted as a location to meet with others. While space is staged and scripted for certain use, there exists an additional form of scripting, as over time, social actors assisted by institutional actors or elites foster a cultural heritage of place which is transgenerationally passed down. In reflecting on the urban space of peace, institutional actors hold power in defining space, and the use of space, this is visible in Mostar through the jurisdictional division of the city along the former conflict lines. The division is made visible in overt and subtle ways from public services to obituary notices in the street,[2] critically these institutional and social signifiers demonstrate the use of space by one or "the other".

This is not exclusive to post-conflict, as the urban is typically a space of 'mixophilia', but also 'mixophobia' defined as spaces of inclusion and spaces segregation, respectively (Bauman 2013: 89). Living in urban areas is notably a 'notoriously ambivalent experience' in that it 'attracts *and* repels' in its anonymity and variety (Bauman 2013: 89). Fundamentally, in divided cities 'mixophobia' can be considered to be promoted by the political elite through the maintained political and educational divides (Bauman 2013: 89). Though the institutional borders in the city are not physical, as Björkdahl (2015: 117) observes, individual social actors in urban environments can -and do- replicate the larger institutional framework. In particular, social actors in divided spaces are educated and socialised to merge their social narrative of movement in line with 'institutional goals', through which social movement replicates the ethno-nationalistic division of the city (Björkdahl 2015: 117). The staging of the city space as divided aims to direct social actors' movements. Notably, this is not just a feature of divided cities but all cities, as cities are sites of everyday conflict due to the milieu of social actors using the space. Fundamentally, as an efficient condensation of social and monetary zones, urban space is a commodity, one which has been instrumentally staged for certain actors. The cost of space in cities and maintaining social interaction in these spaces is demonstrated by the efficiency of urban spaces (shopping centres with cafes and restaurants) (Lefebvre 2009: 53; Ritzer 2012). Engagement with urban spaces, and the perceptibility of those using such spaces to elite manipulation, is often influenced by socio-economic conditions. It is important to note that not all space in urban areas requires money to interact with it, and social actors can move through and use some urban spaces at little cost.[3] While institutional actors stage post-conflict urban spaces, social actors have the agentive capability use urban space in a function alternative to its staging, thereby transforming the use of the space.

SOCIAL (AND PERSONAL) PEACE

The spaces of social and personal peace are important to consider in BiH, while peace may always be considered social, a social peace is enacted on a daily basis between individuals interacting with one and other (Mac Ginty 2013: 287). Social peace is the interaction of individuals, this interaction can rescript conflict narratives and allow for the formation of new or renewed relations which transform the narrative of the conflict. The previously discussed, questioned legitimacy of external involvement, what Björkdahl and Gusic (2013: 317–318) regard as 'friction' between the 'hegemonic discourse' of the liberal peace and local peace, demonstrates the importance of the involvement of social actors in peace processes. It also reflects the potential antagonisms of external institutional definitions of spaces of peace. A key problem with the top-down institutional staging of peace is that it is conceptually, and spatially disconnected from the social use of post-conflict spatialities. While urban peace may be fostered through shared spaces and social movement, it is social actors who perform rescripting and can influence movement in their personal circle of friends and family. In conflict, through displacement and threats to personal safety, which alter the social stage, there is notably a rescripting of social rules and social norms. In particular, during intra-state conflict social roles can be transformed. This generally involves an 'ascribed' status, re-classified as 'master status' (as one aspect of identity that is considered as a variable above other aspects of social identity) (Newman 2013: 27; Brownfield et al. 2001: 73–98). Significantly, the institutional spatialisation of divisions prioritises contingent aspects of an individual's identity. Such narratives seek to create a 'collective memory' to anchor social actors within staged place (Halbwachs 1992: 38; Zerubavel 1997: 4; Zerubavel 2012: 12). As space and identity are intrinsically linked; during an intrastate conflict, an individual's social movements are altered through displacement, mobilisation or due to the threat of violence in everyday spaces. Additionally, social actors may be militarised in intra-state conflict, due to a restaging of the spatiality. For example, the staging of the spatiality of BiH as in a state of war transformed how social actors rescripted space. This can be analysed as such, through the criteria for the mobilisation of '"military-obliged" persons' which constituted all males between the ages of eighteen and fifty-five (UNHCR BiH 1995: iii). While women were not called to perform military service, if between the ages of 'eighteen and

twenty-seven', they were allowed to volunteer services, under the provision that they were 'not pregnant and [did] not have a child under
seven years of age' (UNHCR BiH 1995: iii). Moreover, citizens who
were not obliged to join the military were directed 'to report without
delay to local civil defence units', this included men 'between the ages of
eighteen and sixty-five years...and women between eighteen and fifty-five
years of age' (UNHCR BiH 1995: v). Consequently, during the war age
and gender impacted on the stage that social actors were expected to use.
Furthermore, while women were 'not compelled to join civil defence
units' (UNHCR BiH 1995: v), they were sometimes asked to provide
support in other ways, through the '[preparation of] food or [making]
clothing for soldiers or vulnerable peoples in the community on a more
or less voluntary basis' (UNHCR BiH 1995: v). This directs movement
along traditional patriarchal gender roles, which situate the space of the
home to be feminine, with women assigned to private spaces and a historical marginalisation from public spaces (Tonkiss 2005: 100). Notably,
gender is a variable in the use of all space. An important example of this
is the distinction between the public and private spheres with women traditionally assigned to private spaces. The spatial segregation of women
to the home is reinforced by the impact children had on the capabilities
or movement of women and is important to note, in that the childcare
of children younger than seven, took precedent. Another variable, which
directed the stage an individual was expected to perform on during
the conflict, was the element of being a skilled worker. The Municipal
Secretariat of Defence in BiH had the capability to authorise working
permits (UNHCR BiH 1995: iv) which included 'employment in government offices, public schools, public utility companies, humanitarian organisations, defence industries and strategically important sectors
such as agriculture, mining and transportation' (UNHCR BiH 1995).
Individuals in these positions of employment were issued with a certifi-
cate which meant they avoided conscription for as long as they remained
employed (UNHCR BiH 1995: v). Through the legal policies of BiH
regarding conscription and civil defence units, it can be surmised that the
institutional expectation, of roles to be performed during the conflict,
differed dependent on gender, age and professional skill set. This therefore impacted on the stage that the social actor performed on, directing movement through mobilisation, or maintaining movement due to
skilled worker credentials which would have had a direct impact on the
individual experience of the conflict.

GENDER AND SPACE

Our personal or collective experiences of space direct future use of space. Collective experience may be in the sense of experiencing space with a group of people or from socially translated messages of space, in so far as, we do not need to directly experience a space to form a conceptualisation of it. Furthermore, our movement and use of space is affected by not only those in our social circles or whom we interact with, but also those who we experience space with in space-time. This, of course, differs across time, time in the different hours of the day or night, and in the years of our lives. The physical places we use are space-time dependent, space-time divisions structure our everyday lives and are functional. That noted, the enacted use of space is dependent on the actors using the space but also the time, and the latter may impact on the former. The previously mentioned example of a park demonstrates space-time divisions of movement, parks are popular spaces during the day however, at night, they are when the facility is there, sometimes locked to prevent access, or are avoided spaces. The open space, which during the day is a space of freedom, under different temporal circumstances becomes a space of danger.

This is true more generally in urban spaces, though in establishment, urban spaces became 'a site of agency for women' functioning as a space of increased visibility while holding the potential for invisibility or blending into the crowd (Tonkiss 2005: 100, 102). Furthermore, modern cities have contributed to women experiencing freer movement, as Massey (1994: 258) notes in cities women present a threat to patriarchy due to the relative freedom urban spaces provide;

> in the metropolis we are freer, in spite of all the also-attendant dangers, to escape the rigidity of patriarchal social controls which can be so powerful in a smaller community. However, the increased visibility and movement of women in urban spaces threatened existing order and such spaces presented and continue to present, certain dangers to women which can impact on movement (Massey 1994: 258). The city has always been a space of opportunity in its heterogeneity, this extends our understanding of the city as spaces of intense collaborative scripting, which, dependent on other external conditions, can also create spaces of intense conflict. For example, the opportunity of movement and interaction has not always worked in our favour, as in these spaces we counter dominant power structures and as Wilson noted in (1992: 157);

there is a fear of the city as a realm of uncontrolled and chaotic sexual
licence, and the rigid control of women in cities as been felt necessary to
avert this danger.

Similarly, to what Massey (1994: 258) briefly notes in the previous
quote, city spaces can be particularly dangerous for women. This is a
consistent threat that transcends time and space as Carr (1992: 97) notes
this results in a lack of use and movement in public spaces. However,
different types of insecurity can increase this (Action Aid International
2013). Conflicts can exacerbate this threat and produce different mechanisms to enact violence against women. Sexual violence against women
was one of the defining weapons of the Bosnian war, with more than
'20,000 Muslim girls and women' raped in Bosnia since the start of the
war in April 1992 (UNICEF 1996). Rape exposes women to sexually
transmitted diseases such as HIV/AIDs and has lasting physiological
and psychological implications. Frequently, younger girls are particularly vulnerable to this type of violence this is demonstrated by the tar-
geting of teenage girls in BiH and Croatia, with those who had been
impregnated 'forced to bear "the enemy's"' child (UNICEF 1996). This
also occurred in Rwanda, with adolescent girls who were made pregnant through rape socially stigmatised, as a result 'some abandoned their
babies' while others committed suicide (UNICEF 1996). In Mostar,
rape was used in the process of displacing women from West Mostar by
HVO forces, in particular. The International Criminal Tribunal for the
former Yugoslavia (ICTY) Prlić Judgement IT-04-74-T (29 May 2013:
199–200) notes the frequent use of rape and gang rape including the
rape of a 16-year-old girl. Rape and sexual assault are typically directed
towards women, and intended to intimidate, disempower, and humiliate
the victim. The trauma of such attacks can be long lasting, and the ferocity of the violence contributes to the difficulty of overcoming the transgenerational legacy of the conflict which can also impact on movement
and use of space. This is a particular issue in post-conflict BiH as victims of sexual violence during the war lack adequate support (Amnesty
International 2017). In focus, a lack of support for those who have experienced trauma impacts on their everyday lives and can limit movement,
use of space and mental health and wellbeing.

In intra-state conflict, violence, and in particular sexual violence as
discussed above, occurred in the victims' homes (ICTY Prlić Judgement
IT-04-74-T, 29 May 2013: 199–200). This violence transforms the

experience of the space and may likely be so traumatic that individuals would not want to return to the space due to post traumatic stress disorder. Such violence not only violates the victim but also their spatial sense of the safety of the home. Critically, space is gendered in that, private space is associated with women and public space is associated with men (Massey 1994: 258). Though, this gendering of space does not translate to safety as domestically based violence is frequently directed towards women,[4] reinforcing the threat faced by certain individuals in all spaces.

Critically, the devaluing of women is reflected in post-conflict reconstruction as the narrative of the public divisions takes precedent over private spaces, and private use of public space. Through this we can identify how analysis of post-conflict space is gendered and that this is a replication of the top-down production of space. Massey (1994: 258) notes the gendering of space as a concept in juxtaposition to time;

> where time is dynamism, dislocation and History, and space is stasis, space is coded as female and denigrated. But where space is chaos[...]then time is Order...and space is *still* coded female, only in this context interpreted as threatening.

Space is intrinsically gendered, and public spaces are prioritised as spaces of action over private spaces. Enloe (2014: 34), considers that 'women's collective resistance to [...] feminised expectations can realign both local and international systems of power'. In a post-conflict context, the agentive capabilities of women are underexplored, however, similar to what Enloe (2014) proposes women have a unique sphere of influence. The collective struggle, though differing in regard to positionality—is something that has the potential to unite women across spatialities and indeed through time.

INTERACTIVITY OF THE SPACES OF PEACE

As the conceptual space of institutional peace directly influences urban and social spaces of peace, the maintained institutional segmentation of identity creates distinct social roles of 'in-groups' and 'out-groups', and a manifestation of the institutionally invisible 'other' (Chafetz et al. 1999: 17; Said 1979). Social support of the institutional staging of ethno-nationalistic space presents an issue for the facilitation of the social participation of 'out-groups' which experience a distinct exclusion from society and involvement in political/social spheres of life (Tajfel et al. 1971; Chafetz et al. 1999: 17). In BiH exclusion from society can be traced to elite manipulation of ethnic identity, this is defined by Newman (2014: 28), as the process whereby leaders construct 'ethnic "myths" to serve political interests' these are designed to 'exploit the insecurities felt by people in divided societies in situations of political volatility [...] to promote group coherence, generate resources and mobilise support'. In intrastate conflict, disputes predominately occur over contested land or resources, and in some cases, enlistment is through fear or subterfuge. The conflict is typically fought over the spatiality of the local, and the 'other' is potentially a former neighbour or a family friend. The cleavages of the conflict can be along ethnic or religious lines; which constitutes a prioritisation of 'ascribed' status over 'achieved' (Newman 2013: 27). The prioritisation of aspects of 'ascribed' identity as a motivator of intrastate conflict and can be linked to the enduring myth of 'the other' (Newman 2013: 27; Said 1979). However, while political actors maintain ethno-nationalistic divisions for political means, social actors, who do not holistically benefit from political divisions, have the potential to transform them. The task of the conflict transformation, therefore, is the rescripting of the narrative of the 'other', and the existence of divisive 'in-groups' and 'out-groups' (Lederach 2003: 3; Said 1979; Tajfel et al. 1971; Chafetz et al. 1999: 17). This is a focus of bottom-up peacebuilding which prioritises 'local versions of peace and everyday practices' with an aim to deconstruct the image of the other (Björkdahl and Gusic 2013: 317–318). In post-conflict spaces, social movement across institutional divisions has the potential to transform institutional narratives of divisions through new experiences with other social actors in space.

As observed by Björkdahl and Gusic (2013: 318) the 'social and material divides [in Mostar] originate from or were amplified by a violent conflict as well as by international, liberal peacebuilding efforts'. In focus, international intervention may exacerbate conflict divisions and may also, overall, be incapable of staging a definitive end to the conflict. International involvement can, and has in Mostar, marginalised the capabilities of social actors' which may critically result in limited social movement. Notably, conflict transformation is not limited to one space of peace but is spatially fluid as a concept and impacts on systemic and personal spaces. At a personal level, conflict transformation rescripts perceptions of conflict and larger society; of the safe and the unsafe, removing the dialogical distinctness of 'positive or negative' spaces (Lederach

1995: 7; Galtung 1967: 12). The relationship between the personal and the social is interdependent, as the 'transformation of personal relationships facilitate the transformation of social systems and systemic changes facilitate personal "transformation"'(Lederach 1995: 7). Therefore, the relationship is permeable as social movement rescripts personal experiences of space, but also rescripts the experience of other individuals also using the space, this has transgenerational implications in terms of changing the systemic divisions. Notably, some spaces are more likely to become a space of transformation such as youth centres or schools open to children from all ethnic groups. Nevertheless, all spaces have potential to be a space of transformation in, and following, intrastate conflict, this is somewhat dependent on the individual (and their personal spatial trajectory), and how that interacts with another individual (and their personal spatial trajectory) using the same space.

In conflict, spaces that were once familiar or safe can, through a transformation facilitated by restaging and rescripting, become spaces that are distinctly unsafe and inherently unfamiliar. For example, during the siege of Mostar residential and commercial streets were targeted by sniper fire. These violent actions transformed the narrative of the space, and due to top-down politicising which instigated the conflict, represented the restaging of the space from one of safety to one that is unsafe. Conflict transformation as a process is long term and relationship focused. Our relationships exist in space, we are as spatial as we are social. Indeed, the spatial is how we "know" ourselves, and our social roles. Social peace concerns the stability of social roles, in which social actors operate a conscious informal process of conflict transformation. However, individual social roles, are generally significantly altered during intrastate conflict.

Returning to the influence of institutional actors in instigating the conflict, the mobilisation of civilians is typically directed by institutional actors. Therefore, in stopping the conflict, demilitarisation and demobilisation, can facilitate peace in other spaces due to the interactivity of interrelationship of the spaces. This works both ways, due to the interactivity of the spaces, conflict narratives can also be maintained. As can be seen in Mostar (and BiH at large) the spatial conflict divisions are reflected in the divided education system which presents a major challenge to conflict transformation 'and the long-term stability of a city that is still struggling to reconcile, rebuild and reknit its social fabric' (Björkdahl 2015: 109). Despite such institutional divisions there exists potential for change, with the 'two schools under one roof' ethnically segregated education system 'recently banned' in BiH (Björkdahl 2015: 109). The ban, however, has since been contested and the ongoing issue is explored further in Chapter 5. However, the momentary banning of the system represents a challenge to ethno-nationalistic divides in the country. Fundamentally, the institutional maintenance of ethnic divisions further compounds notions of the 'other', which due to the interactivity of the spaces can permeate to social levels, and through this can establish the narrative of the conflict as intractable (Said 1979).

Through movement, an individual can build peace in their everyday lives, in the shops they frequent, and the streets they choose to walk. Fundamentally, the importance of social involvement in conflict transformation is grounded in the social nature conflict. As such, the effects of conflict on the individual can result in psychological trauma, or traumatic grief which can 'inhibit reconciliation', potentially resulting in a conflict relapse, and perpetuating conflict divisions (Stimec et al. in Mayók and Senehi 2011: 148; Bonanno 2004: 20). The localisation and ferocity of violence in intrastate conflict results in civilians suffering directly or indirectly as a result of death, displacement, torture, mass rape, mass killings and genocide, which were the prevalent forms of violence in the Bosnian War 1992–1995 (Johnson and Thompson 2008: 36; Layne et al. 2001: 277; Errante 2009: 261). Accordingly, conflict transformation emphasises an awareness of the complex dimensions of conflict, and the influence of conflict on the identity and collective memory of generations which follow (Lederach 1995: 7). In particular, Lederach (2003: 3) observes the importance of the personal peace as one of the 'change goals in conflict transformation' (Lederach 2003: 27). Lederach (2003: 27) presents this concept through identifying that conflict inevitably changes personal relationships; but through conflict transformation approaches 'the destructive effects of social conflict' can be minimised, while initiatives can 'maximise the potential for growth and well-being in the person'. The inclusion of the individual actor involved in conflict transformation can be found within the other change goals of 'conflict transformation' as 'relational, structural and cultural', which highlights the interaction of the individual within the wider socio-cultural context (Lederach 2003: 27). Therefore, local social peace initiatives aim to provide a conflict transformation through the acceptance of the conflict, and the transformations achieved through 'mutuality and community' (Lederach 1995: 21). While individuals' can rescript conflict divisions through movement and social interactions in urban space, this capability can be considered to be dependent upon the social roles an individual has previously performed, perhaps through mobilisation, and also their immediate social circle. Through this we can revisit Goffman (1971) and Massey (1994) and the performative nature of social interaction which operates through space-time. Fundamentally, our social performances are reflective of the audience in the space at that particular time, our performance is informed by, but also influences, their performance. Furthermore, the performance of identity is reflective and linked to the socio-psychological effects of individual behaviour. In other words, other social actors influence an individual's thoughts and actions, which, dependent on if the behaviour is observed, may in turn influence others. This is what Massey observes as the fluidity of space, in that we are all acting on our own understandings of space and reacting to others understanding of space (1994: 3–4). Fundamentally, our spatial trajectories influence not only how we feel about and use space, and this is intrinsically informed by our perception, but our understanding and assumptions of others using the space. As observed by Tuan, 'feelings and ideas concerning space and place are extremely complex' and can be difficult to fully understand moment to moment (Tuan 1977: 19). This has both positive and negative implications dependent on the use of space being observed and the observer. As previously noted how we perceive behaviour and use of space, or if we do at all, depends on our own spatial trajectories. Critically, how individuals process this may correlate or clash with institutional narratives of divided city space, according to Bar-Tal (2013: 25);

> shared socio-psychological infrastructure plays a determinative role in the development of the conflict [...] [affecting] how individuals perceive their reality, feel, form attitudes, and act, as well as how a society functions as a whole.

This demonstrates the individual psychological basis of rescripting, as social movement or use of space which transforms personal experience of space and relations within space. Therefore, the impact of the social behaviour on individual behaviour and vice versa is important to reflect on regarding Mostar, but notably also in other divided cities. Within divided cities, some individuals may move in closed spaces, in so far as, their immediate social group or family may not interact across the divide, this then sets a precedent for their own movement which then may influence and increase movement (or may reaffirm stasis) in their own network. Furthermore, social support by individuals (in this example as movement across the divided city) is also of particular importance to the success of 'locally owned' peace (Mac Ginty 2011: 59). According to Mac Ginty (2011: 59), '"locally owned" initiatives' are more likely to succeed as they 'do not rely on costly external resources and can connect with cultural expectations and norms of local communities'. Therein a struggle for influence can be located, between 'external and internal, the state- and non- state actors' with potential divisions between '"the internationals" and the "locals"' within post-conflict areas (Björkdahl and Höglund 2013: 289, 291). It is notable that external classifications of peace by institutional-international and institutional-national actors in post-conflict spatialities, have contributed to solidify spatial narratives of division. Therefore, the physical presence of international actors exists as a 'conflict' though not a physically violent one (Mitchell 2011: 1).

To recap, the four spheres of peace; institutional, urban, personal and social are interactive. Firstly, an institutional peace may provide a 'negative peace' in stopping the conflict, which facilitates other spheres of peace (Galtung 1967: 12). Following a conflict, individual social roles may be transformed from the role of aggressor, victim or potentially a combination of both. 'Social' groups may be formed, or deconstructed, and new groups formed through the urban peace via bottom-up peace-building activities which can establish a 'positive peace' (Galtung 1976: 297–298; Galtung 1967: 12). An important precursor to a conceptual positive peace (Galtung 1967: 12) involves interaction between the institutional, urban, social, and personal spaces of peace. Notably, the 1995 constitution of BiH involves the exclusion of ethnic minorities, while securing the main three ethnicities' (Bosniak, Bosnian Croat, and Bosnian Serbian) access to political participation and governance. The existence of institutionalised ethnicity in BiH can be regarded as maintenance of a 'negative peace', in particular regarding the constitutional exclusion of ethnic minorities (Galtung 1967: 12). This institutional staging of identity maintains the ethnic cleavages of the conflict and reflects the interaction of the institutional with the social and personal, through directing ethnic identity. In Mostar, political narratives are spatialised in the urban context of the city and there is an institutional effort to maintain conflict divisions through fostering ethno-nationalistic divides at social, and personal levels through the divided education system. However, this work presents the agentive capability of social actors to transgress and transform (or rescript) the institutional divide. The success of rescripting can be improved by the co-operative interaction of the spaces of peace, though, institutional divisions (political, education and spatial) currently maintain ethnic cleavages caused by the conflict. As proposed, there exists a division between institutional and social actors in Mostar and BiH, through experiences and interests of the maintenance of the divide. Fundamentally, it is in the interest of institutional actors to maintain political and social divisions. Socially, the benefits of division do not materialise and therefore social actors may, over time, engage with shared spaces or expand movement and through this may rescript their own experiences of space, social relations therein, and ultimately transform the narrative of the space itself.

NOTES

1. The postal services are as follows, Hrvatska Pošta Mostar (Croatian Post Office Mostar) on the West, and BH Pošta (Bosnia and Herzegovina Post) on the East of the city.
2. Obituary notices in Mostar are sometimes taped to street lights and walls, they are sometimes bordered in blue with a crucifix in Croat areas, and bordered in green with the star and crescent in Bosniak areas.
3. Though this is somewhat different in larger cities such as Cape Town, South Africa wherein transport networks and costs can limit movement.
4. 'Women' used to refer to any self-defining woman.

BIBLIOGRAPHY

Action Aid International. (2013, February). *Women and the City II: Combating Violence Against Women and Girls in Urban Public Spaces—The Role of Public Services.* Available from: https://www.actionaid.org.uk/sites/default/files/publications/women_and_the_city.pdf. Accessed Feb 2018.

Amnesty International. (2017). *'We Need Support, Not Pity', Last Change for Justice for Bosnia's Wartime Rape Survivors.* Available from: https://www.amnesty.org/download/Documents/EUR6366792017ENGLISH.PDF. Accessed Feb 2018.

Anderson, B. (2006). *Imagined Communities: Reflections on the Origin and Spread of Nationalism.* London and New York: Verso.

Bar-Tal, D. (2013). *Intractable Conflicts: Socio-Psychological Foundations and Dynamics.* Cambridge: Cambridge University Press.

Bauman, Z. (2013). *Liquid Times: Living in an Age of Uncertainty.* Cambridge: Polity.

Belloni, R. (2001). Civil Society and Peacebuilding in Bosnia and Herzegovina. *Journal of Peace Research, 38*(2), 163–180.

Björkdahl, A. (2015). 'Two Schools Under One Roof' Unification in the Divided City of Mostar. In A. Björkdahl & L. Strömbom (Eds.), *Divided Cities, Governing Diversity*. Lund: Nordic Academic Press.

Björkdahl, A., & Gusic, I. (2013). The Divided City—A Space for Frictional Peacebuilding. *Peacebuilding, 1*(3), 317–333.

Björkdahl, A., & Höglund, K. (2013). Precarious Peacebuilding: Friction in Global–Local Encounters. *Peacebuilding, 1*(3), 289–299.

Björkdahl, A., & Kappler, S. (2017). *Peacebuilding and Spatial Transformation: Peace, Space and Place*. Oxon and New York: Routledge.

Bonanno, G. A. (2004). Loss, Trauma, and Human Resilience: Have We Underestimated the Human Capacity to Thrive After Extremely Aversive Events? *American Psychologist, 59*(1), 20.

Brownfield, D., Sorenson, A. M., & Thompson, K. M. (2001). Gang Membership, Race, and Social Class: A Test of the Group Hazard and Master Status Hypotheses. *Deviant Behaviour, 22*(1). Available from: http://www.tandfonline.com/doi/abs/10.1080/01639620175006581 0#. UuT2PxBFDIU. Accessed 26 Jan 2014.

Calame, J., & Charlesworth, E. (2011). *Divided Cities: Belfast, Beirut, Jerusalem, Mostar and Nicosia*. Philadelphia: University of Pennsylvania Press.

Carr, S. (1992). *Public Space*. Cambridge: Cambridge University Press.

Chafetz, G. R., Spirtas, M., & Frankel, B. (1999). *The Origins of National Interests*. London and Portland: Psychology Press.

Chandler, D. (2010). The Uncritical Critique of 'Liberal Peace'. British International Studies Association. *Review of International Studies*. Available from: https://www.sussex.ac.uk/webteam/gateway/file.php?name=chandler1.pdf&site=12. Accessed 20 Jan 2014.

Enloe, C. (2014). *Bananas, Beaches and Bases: Making Feminist Sense of International Relations*. Los Angeles and London: University of California Press.

Errante, A. (2009). Peace Work as Grief Work in Mozambique and South Africa: Postconflict Communities as Context for Child and Youth Socialisation. *Peace and Conflict: Journal of Peace Psychology, 5*(3), 261–279.

Fanthorpe, R. (2005). On the Limits of Liberal Peace: Chiefs and Democratic Decentralisation in Post-war Sierra Leone. *African Affairs, 105*(418). Available from: http://afraf.oxfordjournals.org/content/105/418/27.short. Accessed 19 Jan 2014.

Foucault, M. (1991). *The Foucault Effect. Studies in Governmentality. With Two Lectures and an Interview with Michel Foucault* (G. Burchell, C. Gordon, & P. Miller, Eds.). Chicago, IL: University of Chicago Press.

Galtung, J. (1967, September). *Theories of Peace: A Synthetic Approach to Peace Thinking*. Oslo: International Peace Research Institute. Available from: https://www.transcend.org/files/Galtung_Book_unpub_Theories_of_ Peace_-_A_Synthetic_Approach_to_Peace_Thinking_1967.pdf. Accessed 25 Nov 2014.

Galtung, J. (1976). Three Approaches to Peace: Peacekeeping, Peacemaking, and Peacebuilding. In *War and Defence: Essays in Peace Research* (Vol. II). Impact of Science on Society 1/2 1976. PRIO Publication No. 25–9. Available from: https://www.galtung-institut.de/wp-content/ uploads/2016/06/galtung_1976_three_approaches_to_peace.pdf. Accessed Apr 2018.

Galtung, J. (1990, August). Cultural Violence. *Journal of Peace Research, 27*(3). Available from: http://www.jstor.org/stable/423472. Accessed June 2016.

Goffman, E. (1971). *The Presentation of Self in Everyday Life*. Harmondsworth: Penguin.

Hague, C. (2005). Planning and Place Identity. In C. Hague & P. Jenkins (Eds.), *Place Identity, Participation and Planning*. London and New York: Routledge.

Halbwachs, M. (1992). *On Collective Memory* (L. A. Coser, Ed., Trans., and with an introduction). Chicago, IL: University of Chicago Press.

Hemmer, B. (1997). Bottom-Up Peace Building in Bosnia. *PARC News* (Spring). Available from: http://www.colorado.edu/conflict/peace/example/hemm6275.htm. Accessed 20 Jan 2014.

International Criminal Tribunal for the Former Yugoslavia. (2013). *Prlić et al. Judgement, Vol. 3, Case No. IT-04-74-T, Nos 764, 1581–3, 1585–6* (pp. 199–200, 459–461). Available from: http://www.icty.org/x/cases/prlic/tjug/ en/130529-3.pdf. Accessed 5 Nov 2015.

International Crisis Group. (2014, July 10). *Bosnia's Future* (Europe Report No. 232). Available from: https://d2071andvip0wj.cloudfront.net/bosnia-s-future.pdf. Accessed 20 Sept 2016.

Johnson, H., & Thompson, A. (2008). The Development and Maintenance of Post-traumatic Stress Disorder (PTSD) in Civilian Adult Survivors of War Trauma and Torture. *Clinical Psychological Review, 28*(1), 36–47.

Kappler, S. (2013). Peacebuilding and Lines of Friction Between Imagined Communities in Bosnia-Herzegovina and South Africa. *Peacebuilding, 1*(3), 349–364.

Kappler, S. (2014). *Local Agency and Peacebuilding: EU and International Engagement in Bosnia-Herzegovina, Cyprus and South Africa. Re-Thinking Peace and Conflict Studies*. Basingstoke: Palgrave Macmillan.

Keil, S., & Perry, V. (2016). Introduction: State-Building and Democratisation in Bosnia and Herzegovina. In S. Kiel & V. Perry (Eds.), *State-Building and Democratisation in Bosnia and Herzegovina.* Surrey and Burlington: Routledge.

Layne, C. M., Pynoos, R. S., Saltzman, W. R., Arslanagić, B., Black, M., Savjak, N., … Houston, R. (2001, December). Trauma/Grief-Focused Group Psychotherapy: School-Based Post-war Intervention with Traumatised Bosnian Adolescents. *Group Dynamics: Theory, Research, and Practice, 5*(4), 277–290.

Lederach, J. P. (1995). *Preparing for Peace: Conflict Transformation Across Cultures.* New York: Syracuse University Press.

Lederach, J. P. (2003). *The Little Book of Conflict Transformation.* Intercourse, PA: Good Books.

Lefebvre, H. (2009). *The Production of Space* (D. Nicholson-Smith, Trans.). Oxford: Blackwell.

Lewis, D. (2010, October 27). The Failure of a Liberal Peace: Sri Lanka's Counter- Insurgency in Global Perspective. *Conflict, Security and Development, 10*(5). Available from: http://www.tandfonline.com/doi/abs/10.1080/14678802.2010.511509#.UuUi5RBFDIU. Accessed 10 Jan 2014.

Lippard, L. R. (1997). *The Lure of the Local: Senses of Place in a Multicentered Society.* New York: The New Press.

MacDonald, C. (2011). Understanding Participatory Action Research: A Qualitative Research Methodology Option. *Canadian Journal of Action Research, 13*(2). Available from: http://journals.nipissingu.ca/index.php/cjar/article/viewFile/37/33. Accessed 20 Dec 2015.

Mac Ginty, R. (2011). *International Peacebuilding and Local Resistance: Hybrid Forms of Peace.* Basingstoke: Palgrave Macmillan.

Mac Ginty, R. (2013). Conclusion. In R. Mac Ginty (Ed.), *Routledge Handbook of Peacebuilding.* Oxon: Routledge.

Massey, D. (1991, June). A Global Sense of Place. *Marxism Today.* Available from: http://banmarchive.org.uk/collections/mt/pdf/91_06_24.pdf. Accessed 20 June 2016.

Massey, D. (1994). *Space, Place and Gender.* Cambridge: Wiley.

McCarthy, J. (2012). *Enacting Participatory Development: Theatre-Based Techniques.* London: Taylor Francis.

Mitchell, A. (2011). *Lost in Transformation: Violent Peace and Peaceful Conflict in Northern Ireland. Re-Thinking Peace and Conflict Studies.* Basingstoke: Palgrave Macmillan.

Neisser, U., & Fivush, R. (1994). *The Remembering Self: Construction and Accuracy in the Self-Narrative.* Cambridge: Cambridge University Press.

Newman, D. M. (2013). *Sociology: Exploring the Architecture of Everyday Life.* Los Angeles: Sage.

Newman, E. (2014). *Understanding Civil Wars: Continuity and Change in Intrastate Conflict.* London and New York: Routledge.

Paris, R. (2010). Saving Liberal Peacebuilding. *Review of International Studies*. Available from: http://www.engagingconflict.it/ec/wp-content/uploads/2012/06/Paris-Saving-Liberal-Peacebuilding.pdf. Accessed 20 Dec 2013.

Richmond, O. P. (2012). *A Post-Liberal Peace*. Abingdon: Routledge.

Richmond, O. P. & Franks, J. (2009). *Liberal Peace Transitions: Between Statebuilding and Peacebuilding*. Edinburgh: Edinburgh University Press.

Ritzer, G. (2012). *The McDonaldisation of Society*. Thousand Oaks, CA: Sage.

Said, E. W. (1979). *Orientalism*. New York: Random House.

Stimec, A., Poitras, J., & Campbell, J. J. (2011). Ripeness, Readiness, and Grief in Conflict Analysis. In T. Matyók, J. Senehi, & S. Byrne (Eds.), *Critical Issues in Peace and Conflict Studies*. Lanham: Lexington Books.

Tajfel, H., Billig, M., Bundy, R. P., & Flament, C. (1971). Social Categorisation and Intergroup Behaviour. *European Journal of Social Psychology, 1*(2), 149–178.

Tonkiss, F. (2005). *Space, the City and Social Theory: Social Relations and Urban Forms*. Cambridge: Polity.

Tuan, Y. F. (1974). *Topophilia: A Study of Environmental Perception, Attitudes and Values*. Englewood Cliffs, NJ: Prentice Hall.

Tuan, Y. F. (1977). *Space and Place: The Perspective of Experience*. Minneapolis: University of Minnesota Press.

UN High Commissioner for Refugees (UNHCR). (1995, February 6). *Situation of Draft Evaders/Deserters from Former Yugoslavia, III*. Bosnia and Herzegovina. Available from: http://www.refworld.org/docid/3ae6b32a10.html. Accessed 22 Oct 2016.

UNICEF. (1996). Sexual Violence as a Weapon of War. The State of the World's Children. *News Feature*. Available from: https://www.unicef.org/sowc96pk/sexviol.htm. Accessed Jan 2018.

Wilson, Elizabeth. (1992). *The Sphinx in the City: Urban Life, the Control of Disorder, and Women*. Berkley and Los Angeles: University of California Press.

Yuval-Davis, N. (2011). *The Politics of Belonging: Intersectional Contestations*. London: Sage.

Zerubavel, Y. (1997) Recovered Roots: Collective Memory and the Making of National Tradition. Chicago, IL: University of Chicago Press.

Zerubavel, E. (2012). Time Maps: Collective Memory and the Social Shape of the Past. Chicago, IL: University of Chicago Press.

CHAPTER 4

Mostar Through Time: *Staging and Scripting in the City*

Providing a historical context to Mostar, to set out spatial (re-)staging and (re-)scripting, is comparable to Ivan Lovrenovićs observations regarding historically unpacking culture; in that those who attempt 'to understand the lines...do not claim to sift the amassed sediment of centuries' (2001: 215). Comparably producing a historical narrative, occurs with specific focus and individual bias. With that in mind, this chapter discusses the transgenerational emergence of present day Mostar, through institutional staging and spatial developments in the city. Mostar, and Bosnia Herzegovina as a whole, has a complex history, some parts of which are better established than others.[1] This chapter does not seek to provide a comprehensive history of BiH as a whole but provides an outline of the spatial staging of the city of Mostar.[2] It is important to note that spatial staging establishes elite narratives or 'invented traditions', these are loosely bound by history and can be used to 'inculcate certain values and norms of behaviour by repetition, which automatically implies continuity of the past' (Hobsbawm 2012: 1). This process is instrumental, and anchors people to place. This is important in the transgenerational narrative of place, as socially and economically, it is people who maintain places. Said (2000: 179) reflects on this functionality of the 'invention of tradition' as a process utilised 'by authorities as an instrument of rule in mass societies' which established bonds between citizens. Therefore, it is important to present the context of contested historical narratives of space and place as, crucially, transgenerational settlement fosters a narrative of spatial belonging. Over time, space

© The Author(s) 2019   71
S. Forde, *Movement as Conflict Transformation*, Rethinking Peace
and Conflict Studies, https://doi.org/10.1007/978-3-319-92660-5_4

becomes staged and scripted as place through transgenerational social movement. Critically, movement and the narrative of the movement affects memory. Said (2000: 178) delves into the complicated matter of 'memory as a social, political, and historical enterprise' which facilitates 'the role of invention'. This is a tool used by elites and is considered to have emerged roughly in 1850; with 'social and political authorities' inventing narratives through the creation of 'such supposedly age-old rituals and objects' for example, 'the Scottish kilt, or in India, the *durbar*' (Said 2000: 178). Through such inventions, Said (2000: 178) observed that elites created a narrative boundary through 'a new sense of identity for ruler and ruled'. Along with differing identities for ruler and ruled, come different spaces, this is identifiable globally. Spaces of political control, are reserved for the elite minority. In intrastate conflicts this takes a more divisive form with spatial displacement and relocation as a part of structural and physical violence which fractures communities (Galtung 1990: 291). The top-down staging of spatialities, seeks to direct social roles and the expected performance of individuals in space. This has a direct impact on the spaces individuals ascribed with these identities are allowed to use or script. Top-down directed narratives were instrumental in directing the Bosnian war 1992–1995, through the coercive narrative of the political elite presenting a formulation of good versus bad and "us" versus "them" which staged the country in conflict and mobilised citizens. Consequently, an understanding of the historical circumstance of narratives of the space is necessary in exploring social, academic, and theoretical narratives of the divided city space.

A Brief History of Mostar

Historically the topographic area of the land and resources provided by these natural spatial forms (e.g. flowing water, shelter, fertile land) has directed human settlement. In Mostar, through archaeological discoveries, human settlement is dated 'since very ancient times' (UNESCO 2005: 17) with the mountainous topography of BiH providing ideal conditions for settlement. As Malcolm (2002: 2) notes, this type of geography contributed to the 'survival of the Basques in the Pyrenees, [and]... the richly-stocked racial museum which is the Caucasus'. Mostar sits between Velež Mountain and Hum Hill, the Neretva River runs through the city. In settlement, the geological formations of mountainous areas provided safety for inhabitants (Malcolm 2002: 2). However, during the 1992–1995 war the higher gradient of the surrounding land generated danger, as offensive measures during the siege of Mostar, by the Hrvatsko vijeće obrane (Croatian Defence Council (HVO)), took advantage of the elevation.[3]

Major spatial transformations in Mostar can be linked back to the periods in which the city was occupied by colonial forces. The Roman Empire conquered the Illyrian land in 9 AD, the movement and settlement around Mostar contributed to the establishment of the city including the construction of a wooden bridge which connected the two banks (Petrovic 2012: 6; Malcolm 2002: 2; UNESCO 2005: 178). During this period the settlement of Mostar was of particular importance due to its proximity to the coast. The area experienced growth and in the 15th century and 'a small settlement began to form around [the] old Roman wooden bridge' (Ravn 1997). Therein, the space was staged as a functional crossing point in the river and a settlement which could be socially scripted by inhabitants. In 1463 the Turkish army conquered the entirety of what was then the 'Kingdom of Bosnia', and an Ottoman feudal system was 'imposed on Bosnia from the outset' (Malcolm 2002: 43, 47). Socially, the quality of life, even for those at a low level of the feudal system, would have constituted improved circumstances when compared with 'feudal pre-Ottoman Bosnia' (Malcolm 2002: 48). Nevertheless, there was a distinction made between 'Ottoman (meaning the entire military-administrative class)' and others (Malcolm 2002: 48). However, the identity of Ottoman was accessible to individuals who adopted an 'Ottoman outlook and Ottoman behaviour' (Malcolm 2002: 48). Through this, Ottomans distinguished themselves from 'raya (flock or herd)'.[4] This was a designation attributed to non-Ottoman inhabitants (Malcolm 2002: 49). As the classification was dependent on the adoption of Ottoman culture, the designation of raya could apply to Muslims also (Malcolm 2002: 49). Indeed, religious autonomy and also the ability to own land was a possibility for all (dependent on 'loyalty' and 'acceptance' of Ottoman ways), however the integration of religion and lifestyle drew closer in the sixteenth and seventeenth centuries (Malcolm 2002: 49). The initial space of Mostar was designated as a settlement, due to the river crossing (UNESCO 2005: 46). On arrival in Mostar the Ottoman forces found a small settlement of '19 houses' between the space of the Roman bridge and Mejdan (a large public square) in the city (UNESCO 2005: 20). The Mejdan was the future centre of the first 'residential micro region' or mahala and the location of the first mosque in Mostar, constructed in 1474 (UNESCO 2005: 23). This was also the first year the name 'Mostar' was designated to the settlement, which encompassed the space 'between the bridge fortification and Mejdan' (UNESCO 2005: 20). The city was a thoroughfare to the Adriatic and underwent further expansion and socio-cultural development as a result of its geographical position (Petrovic 2012: 66). As movement to and through Mostar increased, the Roman 'insecure suspended bridge' was restaged by Ottoman institutional actors, and although at this time still a wooden structure it was more robust than the original Roman bridge (UNESCO 2005: 20). The Ottoman era led to the institutional transformation of the city, which influenced the social use of the space, through the establishment of a new social system, and increased movement to and from Mostar. The city was of geostrategic significance to the region, and occupation of the space aligned with Ottoman goals to dominate 'trade routes, both overland and maritime… across the eastern Mediterranean' (Pamuk 2000: 59). In 1520, the reign of Suleiman II the Magnificent began, following which, construction in the city increased with the 'spatial, constructional and decorative systems, based on [Ottoman] aesthetic principles' (UNESCO 2005: 20). One of the main constructions during the Ottoman era was the bridge, Stari Most ('stari' meaning old and 'most' meaning bridge in Bosnian-Serbo-Croatian or BSC, herein referred to as Old Bridge) which is one of the most well-known Ottoman constructions in the city and indeed the region. The visually striking and practical structure of the Old Bridge transformed the area from a small settlement to a, of the time, 'cosmopolitan crossroads' which eased the movement of produce, and people, and facilitated the growth of Mostar (Architecture Week No. 203 2004: 2). The architecturally celebrated Old Bridge contributed to the development of Ottoman economic and territorial influence. The Ottoman staging of the city space in Mostar also facilitated social movement and the development of public space, and among other Ottoman-influenced cities (such as Sarajevo and Banja Luka) Mostar thrived as a 'strategic river crossing and intersection of trade routes' (Riedlmayer 2002: 103). The Ottoman construction of 'bridges, bazaars, inns for merchants and travellers, and other social service institutions' established the city culturally and commercially (Riedlmayer 2002: 103). As the city was further developed Mostar became 'the centre of Turkish rule' (UNESCO 2005: 178). During the Ottomn occupation, the city was significantly restaged with the construction of 'public baths, hans, thirty mosques, seven medreses, residential quarters…and fortifications' (Yarwood et al.

1999: 1). This increase in spatial development also meant a population growth and in the latter part of the seventeenth century, the population of Mostar had 'reached 10,000' expanding from the initial nineteen houses in the fourteenth century (UNESCO 2005: 20, 21).

From this time until the end of the 19th century Mostar experienced relative ethno-religious peace and stability under the Ottoman Empire. However, though religious diversity was typical of this period, it was socio-economically favourable for individuals to subscribe to Ottoman practices and customs (Malcolm 2002: 43, 47–49). While the occupation had a positive impact on city space through establishing community links and infrastructure, which promoted social scripting of the city space, the occupation also led to a change in demographics in Bosnia at large. This change in demographics is debated within BiH, as 'Conversion vs. Islamisation' (Alibašić 2014: 430–431) which reflects the freely chosen or forced nature of a change in religion during the period of rule. In the context of Ottoman occupation, the influence of the colonial force is evident regarding the social categorisations of Ottoman vs raya (Malcolm 2002: 48–49). While the adoption of Ottoman lifestyle was advantageous to individuals it was not a forced process. Furthermore, during the time of Ottoman occupation, religious co-existence in Mostar can be observed due to the city becoming the 'seat of the Metropolitan (head of an ecclesiastical province) in 1767' (UNESCO 2005: 21). Historical accounts of religious tolerance and diversity can also be spatially located through the building of the first Catholic church in 1847, 'the Bishop's residence at Vukodol' and a cathedral in Podhum in 1866 (UNESCO 2005: 21). However, as the Ottoman Empire declined, and came into conflict with the Austro-Hungarian Empire, unrest spread across the Balkans.

In 1878 at the Congress of Berlin it was decided that Bosnia would be transferrred to Austro-Hungarian administration, this entailed occupation and in 1908 the territory was annexed (Malcolm 2002: 134; UNESCO 2005: 179; Yarwood et al. 1999: 1). The influence of the Austro-Hungarian Empire to physical infrastructure in Mostar is observable through the expansion of the Ottoman cityscape. The new colonial administrators transformed the city establishing 'new layout patterns with larger dimensions...railways and modern roads, public lighting, gas, power, waterworks, industry and capitalist forms of organisation, and advanced education' (Yarwood et al. 1999: 1–2). In Mostar the city council was cooperative in the restaging of the city, as the 'broad avenues and an urban grid' in the West of the city were accompanied by residential and social investments in the city (Pašić 2004: 7). Regarding the location of development of the city, it observable that the Austro-Hungarian government 'saw the city's past and present on the East bank and its future on the West bank of the Neretva' (UNESCO 2005: 24). This is physicallised in the spread of development from the East to the West of the city, a pattern which remains today. Overall, the Austro-Hungarian occupation facilitated an expansion of infrastructure, which contributed to the available public space in the city. The manufactured nature of this merging in the wider city space can be observed in Pašić's (2004: 9) critique of the architecture of Gymnasium Mostar (Gimnazija Mostar—a two schools under one roof school, in Fig. 4.1). Pašić (2004: 9) observes the Gymnasium to be an artificial representation of 'Mostar's Ottoman past', through the utilisation of Islamic aesthetic but in the 'styles of Spain and North Africa'.

This is an example of the typical Austro-Hungarian practice of cultural harmonising in occupied spaces (Pašić 2004: 8). However, as the

Fig. 4.1   Gymnasium Mostar (Photo taken by author in June 2014)

building's Islamic aesthetic stems from the 'Spain and North Africa,' it is disconnected from 'Mostar's Ottoman past' (Pašić 2004: 9). While there is an element of hybridity to such architecture, there are notable issues highlighted by Pašić (2004: 9) regarding the misplaced geographical specificity of the 'Orientalist details'. However, despite such inaccuracies, the city planning, and architectural plans of the Austro-Hungarian administration were not met with resistance but were in fact 'shared and executed by city leaders' (UNESCO 2005: 24). The restaging by the Austro-Hungarian Empire significantly transformed the city space and was characterised by the introduction of larger, multi-storied buildings and 'aligned and symmetrical [centres] where before they had been intimate and delicately varied' (UNESCO 2005: 24).

Critically, this period involved a significant change of the narrative of development taking place in the city. The administration restaged Mostar to replicate Western city spaces, but significantly sought not to interfer with "traditional" social narratives of space. Through administration of the space, the Austro-Hungarian Empire, according to Hungarian diplomat Benjamin von Kállay, was a civilising force which aimed to integrate with Bosnian society (Velikonja 2003: 120). The civilisation narrative of colonial occupying forces is important to reflect on in taking into account the influence of international actors in the space of BiH. While BiH was considered to be resource rich, the rugged topography of the land made it difficult to conquer and, as such, it was of little value, also due to resident 'fractious noble landowners' (Malcolm 2002: 14, 136). Kállay observed BiH to be a markedly 'chaotic country engulfed in, [...] confusion, corruption, and anarchy' (and fundamentally in need of saving), however, significant social and political unrest marked the time of—and attempted to resist—the Austro-Hungarian Empire's occupation (Velikonja 2003: 120). The October 1908, annexation of Bosnia was met with contestation in Serbia, as Bosnian territory was now out of reach (Malcolm 2002: 150). This stoked Serbian nationalism in the early 1900s, and tensions continued during the First World War (WWI) (Malcolm 2002: 153, 154).

It is worth reflecting on the narrative of WWI as one that often starts with the assassination of the Archduke Franz Ferdinand and Duchess Sophie in Sarajevo by a young Serb Gavrilo Princip. The Archduke and Duchess' visit to Sarajevo coincided with the anniversary of the Battle of Kosovo (an anniversary of a Serb military defeat at the hands of the Ottomans). The visit on this day by the Archduke and Duchess is surmised by Malcolm (2002: 20, 155) as an act of 'overwhelming stupidity'. The terms used to describe Princip's actions as revolutionary, or as an act of terrorism, are debated within BiH. As such the weighting of the murders as a contributing factor in the start of WWI varies in narratives, though the assassination is widely considered an important catalyst. Through this start point of the narrative of WWI, top-down (international-institutional) powers stage the inviolability of the colonial occupation. The acceptance of this narrative is evident in Sarajevo, with the purported location of the shooting (now a Museum), hosting a tarpaulin which reads '[t]he street corner that started the 20th century' (Sarajevo Museum sign, June 2014). A picture of Princip and Ferdinand frame the words, with one on either street which meet at the corner. A plaque also marks the point from which Princip 'assassinated the heir to the Austro-Hungarian throne' (Franz Ferdinand memorial plaque, Sarajevo). The starting point of WWI is typically attributed to Princip's violent actions, as the staging of the state of Bosnia as under colonial administration, authorised the visit. This narrative presents the authority of the colonial visit to the occupied space through overlooking the inherent violence of colonial occupation. Tying the start of WWI to Princip's violent actions, and not the violence of Ferdinand's visit, or the occupying violence of the Empire, hypocritically celebrates Western nationalism while problematising Balkan nationalism. This is not to debate if the concept of nationalism is inherently positive or negative, but to highlight the systemic problems of the internationally directed narrative.

Towards the end of the WWI the 'First National Government of Bosnia and Herzegovina' was created as Austro-Hungarian power declined (Malcolm 2002: 162). In 1929 the space became the 'Kingdom of Yugoslavia', under King Alexander, the renaming of the spatiality accompanied a spatial alteration of the territory, which removed 'old regional identities from the map' restaging the heterogenous space (2002: 168–169). The formation of 'nine banovine' constituted the partitioning of Bosnia which had not occurred in over four hundred years (Malcolm 2002: 169).

WORLD WAR TWO TO SFRY

Throughout World War Two (WWII) the space of Yugoslavia experienced multiple conflicts involving various ethnically mixed groups. As Palmberger (2016: 52), observes the three main groups involved in the conflicts, Chetniks, Ustašhe, and Partisans 'cannot be clearly distinguished along national lines' and while the majority of Chetniks and Ustašhe were Serbs and Croats respectively, this was not fully representative of the composition of the groups. Nevertheless, the divisions of Second World War are sometimes replicated along divisions from the 1992–1995 war. During WWII, the city of Mostar was declared part of the Nazi controlled Independent State of Croatia or Nezavisna Država Hrvatska (NDH) and suffered a large loss of life and significant damage to residential buildings. However, damage to the infrastructure in the city was minimised by the Partisans who defended key points of infrastructure and 'had prevented the [planned] destruction of the bridges' (UNESCO 2005: 26). Following the Nazi Germany invasion in 1941, the then hopeful future leader, Josip Broz Tito, rose to prominence through leading the Partisans (Domin 2001a). As a result of massive casualties sustained to citizens of (the then) Yugoslavia, and the atrocities during WWII, a dual historical narrative was imagined (Domin 2001a). The narrative included the unification of the people of Yugoslavia and a top-down prohibition of discussion and public remembrance of the atrocities committed during the war (Domin 2001a). Tito's period of rule led to the conceptualisation of the Socialist Federal Republic of Yugoslavia (SFRY), with BiH granted independence as a republic due to its socio-cultural 'historic existence' (Domin 2001b). Under Tito's totalitarian regime 'strict rules against the expression of "nationalism"' maintained an enforced peace, through the concepts of 'Brotherhood and Unity', this was central to Tito's narrative of the newly formed republic (Domin 2001b). During the time of the SFRY, the state recognised 'Bosnian-Croats and Bosnian-Serbs' as ethnic groups and in '1968, the Bosnian-Muslims were also declared to be a distinct nation' (Domin 2001b). As part of SFRY, Mostar experienced significant growth in residential population from '18,000 to 100,000' (Pašić 2004: 9). Residents of Mostar during this time also benefited from the construction of 'relatively good residential tower blocks', and an influx of locally based industry (Yarwood et al. 1999: 2). The growth of local industry included the establishment of a 'Soko plan', 'Herzegovina Auto' and 'Unis' which produced 'aircraft components', cars, and computers respectively (Yarwood et al. 1999: 2). Alongside this, there was the addition of 'large aluminium, wine, tobacco, joinery, and food processing factories' (Yarwood et al. 1999: 2). Due to the increased industrial employment opportunities for men and women, the cities demographic profile 'broadened dramatically' (Pašić 2004: 9). This period was a time of increased economic activity and institutional restaging of the spatiality. Fundamentally, the increase in local industry contributed to the expansion and development of the city. This period of development is important to note as it reflects the staging of the space as having a direct impact on employment and therefore social movement, in this context, through a population increase. Economically and socially, Mostar was growing and developing, and during the 1970s and 1980s the residential blocks typical of the 'social programmes of the socialist regime', were constructed (Pašić 2004: 9). Following this period of economic prosperity, a city plan to 'preserve the old town of Mostar' led to a renovation project for the Old Town and Old Bridge (Pašić 2004: 9). The renovation of both, enshrined the city as a tourist location, economically this was a benefit to the city which was awarded the 'Aga Khan Award for Architecture in 1986' (Pašić 2004: 9).

While Ottoman development had centred on eastern parts of the city, there was notably a topographical barrier concerning further development in the East which is why further development in the Austro-Hungarian and SFRY time periods spread out into the West of the city. Due to the steep gradient of land to the East of the city, the large residential constructions of the socialist era were predominately on the flatter western side of Mostar (Pašić 2004: 9). This spatial divide of development in the history of Mostar, during a peak of residential and economic growth is important to consider regarding the current development in the city which is largely situated in the West. Fundamentally, the trend of development in the city precedes the divisions of the 1992–1995 conflict and originated due to the Austro-Hungarian development of infrastructure (UNESCO 2005: 24). So due to the unsuitability of land in the East of Mostar, there is a functional purpose to contemporary construction in the West of the city. However, the location of development (and non-development), which correlates roughly with the conflict division lines, has the potential to exacerbate ethno-nationalistic divisions alongside the current political standstill with regard to the absence of local elections.[5]

THE BOSNIAN WAR 1992–1995

The death of Tito in 1980 ended the enforced ethnic harmony; this change in institutional narratives, coupled with continued economic decline led to a stoking of ethnic tensions, and 'scapegoating' as a tactic to shift blame regarding the economic slump (Domin 2001b; Malcolm 2002). The 600th anniversary of the Battle of Kosovo in 1989 did not pass without an institutional display of the importance of 'Serbian' identity (Bar-Tal 2013: 45). In the Gazimestan speech delivered by Slobodan Milošević (at the site of the Battle of Kosovo which marked the start of Ottoman rule) the protection of Serbian ethnicity was emphasised (Bar-Tal 2013: 45). By 1989, Vojvodina and Kosovo were integrated into Serbia, and in the following years Croatia and Slovenia left Yugoslavia (Domin 2001b). In the 1990s as Yugoslavia disintegrated, both Serb and Croat nationalism grew (Bollens 2007: 170). This was characterised by a 'bombardment of misinformation' and the attempted radicalisation of 'the Serb population', with any Croatian government action framed as '"Ustaša" terror' (Malcolm 2002: 217). While ethno-nationalistic tensions were increasingly operationalised, during this time, BiH remained varied in its demographics and the Bosnian government attempted to pacify its Serb population during this time of heightened nationalist rhetoric (Malcolm 2002: 217).

In 1991 urban Mostar's population was '29 per cent Croats, 34 per cent Muslims, 19 per cent Serbs, 15 percent Yugoslavs and three percent other groups' (other groups are made up of 'the 17 national minorities') (Yarwood et al. 1999: 2; Bosnia—Herzegovina Council of Ministers 2004: 3). Furthermore, not discounting the 'Donja Mahala' and 'Zahum' areas; as largely Muslim ('60 per cent') and Croat (of a similar majority) Mostar was ethnically diverse with no geospatial ethnicised areas (Yarwood et al. 1999: 3). As such Mostar was widely considered to be a 'a paradigm of a harmonious multi-ethnic society', wherein different religions lived alongside one and other (Yarwood et al. 1999: 3). It was less than a year after the 1991 census that the city was transformed through violent ethno-nationalistic political narratives. Following the independence of Slovenia and Croatia; BiH also sought to leave SFRY. On the 2nd of March 1992 the result of the independence referendum was declared. Through this, a majority of '64% of the Bosnian electorate' voted to establish 'a state "of equal citizens and nations of Muslims, Serbs, Croats and other"' (Domin 2001b; Pašić 2004: 9). The referendum was a turning point in the history of the country and was followed by an increase in violence, with frequent episodes of 'Serb paramilitary forces [...] bombing and shooting in towns throughout Bosnia' (Domin 2001b). The subsequent acknowledgement of the independence of BiH by the European Community on the 6th of April 1992, led to Serb forces firing on 'a crowd of peaceful demonstrators' in Sarajevo (Domin 2001b). Notably, the recognition of the social majority consensus on independence by international elites ignited the full outbreak of violence. Through this the power of social actors is demonstrated via the referendum, but dually so the relevancy of top-down institutional actors (in this example the international community) in the legitimisation of the independent status of BiH. Critically, this reflects the ability of social actors to rescript definitions of space, but also highlights the relevance of the support of institutional actors, which facilitated the referendum and restaged the spatiality (in this latter example, the definition of the spatiality as an independent BiH).

The cultural-conceptual, and physical spatiality of the former SFRY was frequently 'described as the "other" of Europe' from this Todorova coined the term 'Balkanism' to highlight the unique positionality of the Balkans, and the physical and cultural 'space' which it represents. It was this unique cultural space that provided the stage for the 1992–1995 war (Todorova 1997: 3–20, 2009: 3). Todorova explores the lineage of this designation of the spatiality noting that it was named for 'Haemus', the mountain range that became known as 'Balkan, which meant difficult mountain' (1997: 25). This not only reflects the previously noted challenges the topography presented for colonising forces but becomes a useful metaphor in exploring the international attitudes towards the area during the conflict. Following the dissolution of Yugoslavia, the term Balkan outlined cultural boundaries as well as physical, with 'the Balkan spectre' defined as 'not a character but a name, a signifier' with layers of meaning (Todorova 1997: 21). While the word '"Balkan" was being accepted and widely used as a geographic signifier' the layers of the identity imbued in the name was at the same time 'becoming saturated with a social and cultural meaning' (Todorova 1997: 21). The Balkans can be therefore observed in some ways as a 'imaginative geography', albeit one that was "disintegrating" in conflict (Said 1979: 49).

Mostar—Two Wars, One City

Mostar during the wider Bosnian war was the location wherein 'two armed conflicts' played out (Petrovic 2012: 71). The first, in 1992, which lasted April to June was fought by 'Croat and Muslim forces' against the 'Serb-led YA [Yugoslavian Army, JNA] and other Serbian forces' (Petrovic 2012: 71). Within the territory held at the time by

Serbian forces, the 'left bank of the Neretva in Mostar and [the] South of Mostar', ethnic cleansing was directed at the Muslim population (Petrovic 2012: 72). When the Serbian forces retreated from the spatiality of Mostar they left behind a significantly damaged city space (Petrovic 2012: 71). As noted by Petrovic (2012: 72) the 'cultural cleansing [...] brought about economic cleansing' as the 'industry and tourism' cultivated during SFRY were significantly affected by the conflict (Petrovic 2012: 71–72). The impact of the economy on movement is further discussed by Bose (2002: 106) who highlights the mass movement of 'Mostar's intelligentsia and middle-class professionals' during the conflict. It is in this context that the economic impact of the war can be observed, not only in the city but as a variable of who was able to leave the conflict area. Moreover, the ethno-nationalistic violence and 'psychological warfare' pervasive in the conflict, resulted in a significant reduction in the population of Mostar from '120,000 to 30,000' (Petrovic 2012: 72). Notably, fighting between the 'Croatian HVO and the predominately Muslim' (Armija Republike Bosne i Hercegovine (Army of the Republic of Bosnia and Herzegovina (ABiH)) had started before the siege of Mostar, which began on the 9th of May 1993 (Petrovic 2012: 74). The siege initially involved the widespread 'expulsion of Muslims from the right bank' alongside 'ethnic and cultural cleansing' also directed at the Muslim populace (Petrovic 2012: 74).

Notably, how an individual experienced the conflict depended very much on their identity and space of residence in the city, for example, there was considerably more damage in 'East Mostar and the Bosniak part of West Mostar' during the 1992–1995 conflict (Bollens 2007: 171). As Bollens (2007: 171) observes 'between 60 and 75 percent of buildings were destroyed or severely damaged' in these areas. Contrastingly, a significantly smaller 20%, of buildings were destroyed on the West side 'with most [of this] destruction concentrated along the western side of the Boulevar line of hostilities' Bollens (2007: 171). The impact of the conflict on the infrastructure of the city, included the destruction of the bridges and in particular, the Old Bridge. On November 9th, 1993, sustained shelling, attributed to the HVO, led to the collapse of the Old Bridge which had previously stood since construction in 1566–1567 (Calame and Charlesworth 2011: 116). Petrovic (2012: 76) observes the social importance of the bridge, highlighting that though it had taken 'centuries of love and admiration to make this mass of stones more special than many others' the destruction of the bridge took 'only minutes'. The Old Bridge can be regarded as a social monument, as the bridge led to the naming of the city, in this, the bridge is not only a namesake but important to the cultural heritage of the city for Mostarians. The social impact of the loss of the bridge during the Bosnian war (1992–1995), is further exemplified by the description of the impact of the destruction, by a Croatian journalist, Slavenka Drakulić (1993);

> We expect people to die. We count on our own lives to end. The destruction of a monument to civilisation is something else... Because it was the product of both individual creativity and collective experience, it transcended our individual destiny.

The above quote demonstrates the importance of physical space to social narratives of identity. The structure of the Old Bridge, regarded as an of-its-time 'technological wonder', fostered settlement and the social scripting of community in the surrounding physical space (UNESCO 2005: 5). Fundamentally, the Old Bridge is integral to Mostarian identity. This was demonstrated by the prioritisation of the post-war reconstruction of the bridge by local, national and international actors. The bridge was reconstructed and re-opened in 2004, this was 'staged' by institutional actors as the reconciliation of the city, in the same year that the city statute established the institutional divisions of Mostar (Forde 2016: 477). While staged as a symbol of reconciliation, the contemporary usage of the bridge spans various functions (commemorations, celebrations, performances and protests) and highlights the capability of social actors to rescript narratives of space.[6] In short, while the narrative of the bridge as healing the community may overstate the importance of the space, the reconstruction of the bridge was an important spatial transformation in the city and has contributed to the revitalisation of the city through the reconstruction of the transgenerational landmark.

MOVEMENT DURING THE 1992–1995 WAR

In the 1992–1995 Bosnian war, more than a million people were displaced directly by the conflict or were affected by the conflict (Internal Displacement Monitoring Centre (IDMC) 2013). This may have involved, imprisonment, abduction, enlistment, or displacement through fleeing the conflict. In focus, different variables of identity such as gender, age, and ethnicity impacted on people's experiences of the conflict though violence was experienced by all ethnic groups. Though the research did not focus on narratives of the war, some participants wanted to tell their story of the space, participant K is a Bosniak living in Mostar;

> I was born and grew up here [in Mostar], we were refugees, spent one month in a war camp, my mother had to walk ten kilometers on foot [carrying me], and then to Konjic twenty kilometers from there, there was two thousand families [also with us]. After seventy three days in the mountains of Sarajevo at 3am my father returned, he had a big beard and had lost ten kilograms [of weight], we thought the Chetniks [a name for a Serb military group] had come to kill us. (K Interview 2015)

Fundamentally, displacement was widespread during the war, which was tactical in ethnically "cleansing" the space, those targeted depended on the location, but this was largely directed towards Bosniak Muslims. Ethnic cleansing and genocidal rape was used to displace populations and establish spatial ownership of land, this included killing men and boys, and raping women and girls alongside displacing populaces from their homes. One of the most intensive occurrences of violence occurred in Srebrenica in 1993. The town was demilitarised and by May, the United Nations Protection Force (UNPROFOR) had established Srebrenica as a 'safe area', however, subsequent attacks by the Army of the Serb Republic (Vojska Republike Srpske (VSP)) led to a sustained campaign of ethnic-cleansing directed at unarmed civilians (Human Rights Watch 1995: 8, 10). The way this was handled by UNPROFOR has been widely criticised and was also discussed by K;

> UNPROFOR could have done a lot more... 8720 killed the biggest genocide since WWII [...] I am Muslim, every day I ask myself; how can someone be so inhumane. (Interview 2015)

The reported daily consciousness of the Srebrenica genocide for K, demonstrates the transgenetional impact of the massacre, committed by the VSP. This is further evident regarding a lack of a reconciliation surrounding the Srebrenica massacre, which impacts upon the way in which the massacre is remembered, and memorialised. The combined scale of the violence, and a lack of justice for the victims of the massacre presents a significant challenge in post-conflict peacebuilding. Fundamentally, the institutional authorisation of mass violence, combined with the institutional failings of UNPROFOR, led to the genocide of the residents of Srebrenica and 'the worst crime on European soil since the Second World War' (Annan 2005). Critically, transgenerational narratives of violence committed during the conflict (in this case the violence committed toward Bosniaks whom K self-identified as) permeates everyday life for some individuals and therefore may impact on the progress of conflict transformation. Significantly, this narrative demonstrates the interaction between the actors of rescripting and restaging. Through the act of genocide, Srebrenica, was restaged from a "safe area" to a site of genocide, and through a lack of institutional acknowledgement, to a site of injustice, wherein institutional political narratives struggle over the spatiality each year. As Selimovic (2016: 54) observes due to new discoveries of victims, each year there is a 'highly charged commemoration ritual' which goes via Sarajevo 'before continuing to Srebrenica'. Through, participant K's narration of a daily consciousness of the violence committed against fellow Muslims; it is observable that the genocide has a impact on individuals across space but also time. As outlined through K's narrative of his own and his family's movement during the conflict, for many, movement and use of space was limited by institutional divisions. For C, a Croatian citizen, who was a young adult at the time of the war, their experience of the city and of institutional divisions was different. As a result of C's English language skills, she was able to help people fill out the required paperwork to leave the city during the conflict and was also able to leave the city herself (Interview 2015). This represents an important international dynamic at play in Mostar during the conflict, due to the working language of the paperwork, which inevitably would have marginalised non-English literate individuals attempting to move away from the space of the conflict. C also experienced free movement through the city during the war, at a time when movement was restricted. To enable movement, participant C was given a piece of paper by one of the Spanish United Nations commanders saying '"C" can go anywhere in Mostar, signed Jesus' (Interview 2015). What is relevant to consider in this narrative of movement, is that the institutional authorisation restaged the spatiality and restricted movement in the city. However, the skilled position C held; as someone who is literate in English, and therefore someone who was able to fill out paperwork, facilitated her free movement in the city. In reflecting on the staging of the spatiality these narratives highlight that multiple staging's may be ongoing in one particular space, and that this is dependent on the individual, their skill set and institutional recognition. Fundamentally, these variables direct the stages accessible for social actors to rescript or transform.

THE 'END' OF THE CONFLICT

In March 1994, the Washington Agreement 'established a Federation of Bosniaks and Croats' and brought a cessation to military conflict in Mostar, (ICG Balkans Report No. 90, 2000: i). However, as observed by Björkdahl (2015: 113) the Washington Agreement, though bringing an end to the physical violence, entrenched the division in Mostar. This was followed by the 1994 April 6th signing of a Memorandum of Understanding (MOU) in Geneva by 'Croat and Bosniak politicians' (ICG Balkans Report No. 90, 2000: 4). The MOU facilitated the European Union Administration of the city (EUAM) attempted to set the conditions for co-operation, and the reunification of the city, with the respective parties agreeing to assist the EUAM 'to normalise life in Mostar' (ICG Balkans Report No. 90, 2000: 4). A subsequent MOU was agreed upon in July, which set out the particulars of the 'aims and objectives' of the administration (EUAM 1996). This MOU was followed by the signing of the Madrid Agreement on the 2nd of October 1995, which put forward the importance of the memorandum being fully implemented (ICG 2000 No. 90: 7). Following this agreement, the two mayors of Mostar, Safet Oručević (the mayor of East Mostar), Mijo Brajković (the mayor of West Mostar), and Hans Koschnick, (the then EU administrator of Mostar), established an agreement for a governing structure for the city (ICG 2000 No. 90: 7). This was integrated into an annexe to the Dayton Agreement, which would end the wider conflict at an institutional level, and stage BiH as no longer in "conflict". In December 1995, the Dayton Agreement was signed ending the military conflict across BiH. However, this agreement instilled a two-prong power structure which replicated the divides of the conflict, enshrining the divisions politically, and spatially. The Dayton Agreement also set out the spatial divide in Mostar through the establishment of the six municipalities within the city (Annex to the Dayton Agreement 1995: 1). The format of the agreement was characteristic of international intervention in the post-cold war era, guided by the 'United Nation's Agendas on Peace, Development and Democratisation' and shaped by the idea of 'long term involvement' of such international institutions in post-conflict spaces (Chandler 2000: 34, 35). The map of the Balkans was reimagined once more through the agreement which established peace through the ethnically distinct spaces of the Republika Srpska (RS—predominately ethnically Serbian) and the Federation of Bosnia-Herzegovina (FBiH—predominately ethnically Bosniak and Bosnian Croat). Notably, the political structures of the two entities differ significantly. The RS includes 'two levels of government...entity level and municipality level'; while the FBiH has 'municipality, Canton and entity' levels (Živanović 2015: 210). Through the differing political structures, there are notable inconstancies in governance and political responsibilities. For example, the RS has municipalities with 'local mayors and officials...directly elected by local voters', while the FBiH consists of ten cantons with 'each canton composed of municipalities' (Boyle 2014: 106). The FBiH is decentralised 'with powers devolved to [the] 10 cantons', while the RS is centralised (Keil and Perry 2016: 5). There is then the territory of the Brčko District existing as 'territory simultaneously part of the Federation and the RS' (Perry 2006: 56). Brčko sits between the two entities but is self-governing. The district prospered post-conflict as international funding flooded into the city, though ethno-nationalism is still visible in the city, along with divisions in education (Geohegan 2014).

The responsibilities regarding an interim statute for the city 'including support for the unity of the city under an interim structural agreement' (Winterstein 2003: 50) was set out in the Dayton Agreement, which included a call 'for full implementation of the EUAM MOU'. As there was inevitable contestation over the post-war use and division of space in the city, the re-negotiation of city areas was approached strategically. In particular, the establishment of the previously mentioned central zone was a purposeful creation of space, designed to demonstrate the liveability of a shared space in the city (Bollens 2007: 168). Local politicians contested the size and location of this shared space, as observed by Yarwood et al. (1999: 30) 'the Croats made emotional speeches about who had died on which street and how it must [therefore] be forever part of their homeland'. Furthermore, Croat politicians wanted the central zone to occupy a much smaller space than the Muslims (Yarwood et al. 1999: 30). In early February 1996, Hans Koschnick set out the space of the central zone and the six municipalities, which was contested by the Croat mayor Mijo Brajković (ICG 2000 No. 90: 9). The proposed central zone would have included residential space which would have encroached 'far into Croat territory', according to Yarwood et al.

(1999: 31) this 'outraged the Croats' but was accepted by the Muslim politicians. Over the local radio, Brajković announced a dissolution of discussions with the EU and called for a 'demonstration in front of the EU office at Hotel Ero' (ICG 2000, No. 90: 9). This resulted in Koschnick trapped in his car outside the hotel as protesters mobbed the car 'threatening to lynch him' while the 'Croat police stood by and did nothing' (ICG 2000, No. 90: 9). Though the conflict had ended, significant tensions over the ownership of space remained, following this incident a summit in Rome was called.

The subsequent Rome Agreement in 1996 considered the unification of the city. This agreement proposed that the central zone should be a space for the public, specifically for the facilitation of 'local cultural activities' (IFOR, 18th February 1998). Though intended as a collaborative space, the vacuum of defined institutional ownership of the zone resulted in contestation of the ownership of the space instead of collaboration (Bollens 2007: 168). According to Siani-Davies (2004: 78) who interviewed local United Nations workers in 2000, there was notably a lack of both cooperation and communication between the two sides of the 'multi-ethnic police service in the city'. Critically, the central zone became a stage for continued tensions to be played out, as the zone itself was to be 'administered jointly by both sides', this space became a 'no man's land, and a highly contested space' (Vetters 2013: 27). The Rome Agreement 'also established the return of refugees and set out how property matters (the return of residential property) would be handled (IFOR, 18th February 1998). In 1996, the EU administration of the city ended with subsequent arrangements for support to include a Special Envoy to support and promote a range of activities including; the return of displaced persons and refugees, reconstruction, human rights [and] law enforcement (Hill and Smith 2002: 372). The arrangements also included the 'stabilisation and strengthening of the newly elected unified administration [...] freedom of movement' and support in the unification of an 'effective law enforcement system' (Hill and Smith 2002: 372). This was to be overseen by the High Representative and Office of the High Representative (OHR) which was set out in the Dayton Agreement to essentially ensure that the peace settlement was implemented (UN 1995: 112). As noted by Willigen (2013: 70), Annex 10 of the Dayton Agreement allowed for a full range of movement for the High Representative and OHR in the interpretation of their 'own authorities and powers'. This freedom was reflected through the Bonn

Powers, which, as Willigen (2013: 73) observes, 'have been widely used by the High Representative as an instrument for imposing the peace implementation process'. In Mostar, the spatial conflict divisions were mirrored institutionally during the war, this is observed by Calame and Charlesworth (2011: 120) as continuing to function 'separately for several years' following the establishment of the Dayton Accords. Additionally, in the report 'Commission for Reforming the City of Mostar' Winterstein (2003: 13) notes that through this power structure 'the City of Mostar has never come to life'. Fundamentally, the divide of power had resulted in the divisive administrative structures and those in charge assuming authority 'solely for the good of "their own people"' (Winterstein 2003: 13). Notably, international involvement, which successfully legitimised the division of the country, unsuccessfully attempted to force shared pockets of spatial co-existence in the city and country.

POST-CONFLICT DISPLACEMENT AND DEMOGRAPHICS

The 1992–1995 war produced significant displacement within, from, and to the city following the conflict. When the 'hostilities' had ended in 1994 as Bollens (2007: 170) notes 'the demographic and physical composition of the city had been severely reconfigured'. In the post-war period, there was a high number of internally displaced persons in Mostar, with an estimated '30,000' displaced in East Mostar and '17,000' displaced in West Mostar (Bollens 2007: 171). Displaced persons originated from across BiH, with individuals coming from 'Eastern Herzegovina, Stolac and Čapljina [and] West Mostar' living in the East of Mostar; and individuals coming from 'Central Bosnia, Sarajevo, Jablanica and Konjic' living in the West of Mostar (Bollens 2007: 171). The displacement was ethno-nationalistically centred through the politically motivated subsidisation of resettlement. This process was funded by the Croatian government which, according to Bollens (2007: 171), aimed to 'strengthen Croat demographic and political control over the city's destiny'. This demonstrates the relevancy of social actors in space, and that through subsidised resettlement of social actors to live in space, the narrative of ownership of the space can be reinforced.

The return of previously displaced individuals represented a significant challenge in post-conflict Mostar. As Chandler (2006: 124) notes, the number of returnees in Mostar was 'well below the BiH-wide average' in the years following Dayton, furthermore, citizens in Mostar 'continued to be expelled after the end of the war'. However, as with all cities, demographics are not in stasis as the return of Bosniaks to 'Western Mostar' and Serbs 'to both halves of the town' in the early 2000's demonstrates (Chandler 2006: 124). However, Chandler (2006: 124) also highlights the directed influx of residents 'which resulted in West Mostar becoming nearly exclusively Croat, with most Serbs living in nearby regions in the RS, and Bosniaks living in East Mostar'. Writing in 2002 (two years prior to the reunification of Mostar through the 2004 city statute) Bose (2002: 106) observed, that the demographics 'in East Mostar (and to a lesser degree in the western part) consists of dispossessed people, less educated and trained, typically of rural or small-town origins.' Around the same time, in 2003 The International Crisis Group produced an ethno-nationalistic demographic breakdown of Mostar and reported that statistically, the West side of the city was predominately ethnic Croat, and the East side was predominately ethnic Bosniak;

[t]he "ethnic purity" of the electorate of the municipalities ranges from a high of 98 percent (in the Bosniak-majority Municipality South-East) to a low of 77 per cent (in the Croat majority Municipality South). Only the sparsely populated Central Zone can boast of having a "minority" voting population of more than 25 percent. (ICG, No. 150 2003: 7)

Fundamentally, political actors sought to establish a social narrative of the city through the movement of persons which aligned with the conflict division. Through this the societal impact of the institutional staging of the city as divided can be observed, during and after the conflict, as political parties sought to cement territorial gains, legitimised through the Dayton Agreement. What can also be identified is the importance of social actors, who were purposefully resettled in order to legitimise the claim to the space.

MOSTAR 2004

In 2004, the City Statute was implemented by, then High Representative, Paddy Ashdown as part of the reunification of the city. While the statute "unified" the city and attempted to 'neutralise ethno-national supremacy' it notably did not solve the political impasse (Björkdahl 2015: 113). Modifications to the statute included the restaging of the municipalities into city areas (or electoral constituencies) though these changed by name, spatially, the areas corresponded with the previous municipality zones (Ashdown 2004). As noted in the 2004 City Statute, the previous municipality divisions 'failed to unify the City of Mostar' and instead resulted in the creation of 'parallel institutions' which maintained divisions in the city (Ashdown 2004). The 2004 statute also removed the central zone as a unique zone of administration in the city. Reflecting on the city statute demarcation lines, in 2003, Winterstein observed that the lines 'primarily served to mark the "achievements"' of space gained during the conflict (2003: 12–13). From this it can be noted that the maintenance of such division lines, by whatever name, reaffirms the narrative of the conflict. While the city is spatially divided, it is also politically stagnated. The city has not had a mayoral election since 2008 and as a result, Ljubo Bešlić, a Croat politician has held office since 2004. The lack of electoral participation is a concern for many residents. For some participants, this evokes Franjo Tuđjman, the Croatian President's, war declaration that the city of Mostar was to be established as 'a "Croatian Republic of Herceg-Bosna,"' (Bollens 2007: 170). Spatially, the city area divisions set the stage for the local actors to perform on, the city area lines exist physically through the organisation and administration of the city, and in the divided education system. Although the division is not absolute, it is due to institutional divisions that non-movement has the potential to be fostered. However, a complexity of social divisions can be observed in the city (as in many cities). While there are divisions between the three ethnic groups of Bosniak, Serb, and Croat, there are also sub-divisions which may influence movement over and above ethnicity (such as ideology, class, gender, and age).

In Mostar, at the time of the implementation of the 2004 City Statute, numerous spaces were restaged in the city, two important restaged spaces included, the reconstruction of the Old Bridge and the establishment of the OKC Abrašević cultural centre. Chronologically the reconstruction of the Old Bridge came first. The international-institutional involvement of the United Nations Educational, Scientific and Cultural Organisation (UNESCO) led the reconstruction which demonstrates the capability of international-institutional actors to restage post-conflict space. Notably, in 1945 UNESCO was founded as 'political and economic agreements' were considered 'not enough to build a lasting peace' (UNESCO 2013). Adaptively, UNESCO's remit now involves the preservation of cultural heritage, which has social but also socio-spatial characteristics as;

> physical artefacts and intangible attributes of a group or society that are inherited from past generations, maintained in the present and bestowed for the benefit of future generations. (UNESCO 2016)

The importance of such cultural heritage is highlighted by the efforts of UNESCO to 'protect and secure [...] cultural treasures- held in sites, in museums, cultural institutions and [importantly] within people themselves' (Stone and Bajjaly 2008: 194). Though an international-institutional actor, the established mission of UNESCO observes that lasting peace transcends institutional arrangements and is a social and even individually centred process (UNESCO 2013). It is through understanding other cultures and differences, that people can no longer be generalised under a reductive stereotype. The prioritisation of social involvement in the preservation of cultural heritage in post-conflict spaces has important implications for stability and peace. As observed by Stone and Bajjaly (2008: 194–195) in Iraq, which suffered museum looting, UNESCO 'was the first international organisation...to try to use Iraq's cultural heritage' to help bring stability. This is comparable to the efforts to reconstruct the Old Bridge, the bridge was important transgenerationally, it was an Ottoman construction but also, over time, the namesake of the city, a tourist site, and an established landmark. The task of rebuilding the Old Bridge, was not then to produce an identical bridge, but the reconstruction and restoration was to remain 'faithful to the ideas and principles of the original structure' (UNESCO 2005: 7). In the nomination document, UNESCO note the unique 'historical stigmata and...patina' of the Old Bridge (UNESCO 2005: 7). This is something achieved over a long time, while the bridge, at the time of construction, was institutionally constructed and directed by capital, socially and transgenerationally it has been rescripted into public spatial identity. This transgenerational rescripting meant that the bridge as it was, could never be reconstructed, while this may seem obvious to state, it is important to reiterate as the social use transformed the space.

As noted, while an exact replica was not the aim of the reconstruction, traditional methods and of stone cutting were utilised for 'technical, aesthetic and ethical advantages' (UNESCO 2005: 7). Through the work of replicating the bridge, the importance of the visibility of 'cultural heritage' can be observed (UNESCO 2016). As the bridge is of symbolic and transgenerational importance in the city of Mostar, the reconstruction and re-opening were praised internationally as a of reunification (UNESCO 2005: 1). While considered symbolic of the reunification of the city, this does not translate to the lived reality of the space, and the bridge can be considered a 'stage of memory' whereby an institutional narrative of the history of the space is maintained (Calame and Pašić 2009; Forde 2016). The reconstruction of the Old Bridge occurring in the same year that the city statute reaffirmed the divisons of the city demonstrates the incongruence of institutional narratives of peace.

The year 2004 was a busy one for spatial transformation in the city, and alongside the reopening of the Old Bridge and the new statute, the OKC Abrašević (herein where appropriate Abrašević) cultural centre was opened. The youth cultural centre is situated in the former central zone, on Alekse Šantića a former frontline of the conflict, and is a space open to all in the city. While the bridge and cultural centre are distinctly different spaces, the staging and scripting of the spaces in the post-war city are comparable. In particular, it is significant that the reconstruction of the Old Bridge was staged internationally as a 'symbol of reconciliation' (UNESCO 2005: 1). However, while the cultural centre established physical space for reconciliation, the establishment of the centre received little attention. As noted by others, the centre offers a space wherein the narrative of ethnic identity is not a variable in the use of the space (Kappler 2014), though politically it is left-wing (Carabelli 2018: 131). The extent to which it can be considered open to all is debateable due to the socialist origins of the centre. The space which is now Abrašević was, once a socialist workers club, and is named after a young socialist poet, Kosta Abrašević. One of the current purposes of the centre is focused around searching 'for alternative visions and models of the world and society', though no longer a socialist workers club, the political principles remain at the centre (OKC Abrašević 2016). Therefore, ideologically the space may not appeal to all. This is not to critique the work Abrašević does, nor the space, or indeed, the ideology of the centre, but to note that this may deter movement for some. As observed by Björkdahl and Gusic (2016: 95) the centre operates outside formalised peacebuilding processes, and it has been able to resist both 'misdirected international peacebuilding initiatives and ethno-nationalist governance'. Furthermore, the spatial impact of the centre in its location on a former frontline transformed the derelict street and has increased movement and use of the space through the scripting of the centre. While both the

Old Bridge and OKC Abrašević are significant spaces in the city, fundamentally, OKC Abrašević is a socially rescripted space, which provides a unique space for engagement and social interaction in the city, both the bridge and the centre are discussed further in Chapter 6.

CITY ADMINISTRATION

In April 2004, a '"General Concept" for an EU-led mission' was put forward which included the deployment of an EU peacekeeping force (EUFOR) of 7000 to BiH to replace the NATO (North Atlantic Treaty Organisation) force (Kim 2006: 4). The deployment of these forces was supported by the United Nations (UN) Security Council and was to begin in December 2004. The operation named Althea was led from three key areas in BiH; Mostar, Tuzla, and Banja Luka (Kim 2006: 5). The subsequent transfer of power from the Office of the High Representative to elected persons in BiH began with the '[Stabilisation and Association Agreement] SAA negotiations process' in 2005 (OHR 2005). According to the OHR South, the City Administration and City Council in Mostar, had all the necessary 'tools at their disposal to continue the unification process and to provide effective city services to the citizens' (OHR 2005). Though the OHR also maintained that they would continue support for the administration and council in this capacity and in 2006 the promise of support was followed by the OHR establishment of the Spatial Planning Institute, to assist with urban planning in the city (OHR 2005; Djurasovic 2016: 142). Though this appeared to be a unified plan, the process in practice was dominated by fractured attempts at planning the cityscape. As Djurasovic (2016: 42) observes, there was a prioritisation of 'regulatory plans for individual city zones' which corresponded with the six divided electoral areas. The implementation established a fractured approach to city planning and was fundamentally without a city-wide cohesive plan (Djurasovic 2016: 42).

This divisive approach is comparable with ethno-nationalist political tensions. As a result of 16 unsuccessful efforts to elect a mayor since 2008 the OHR intervened to facilitate the election of a mayor 'without a qualified majority' (Björkdahl and Gusic 2016: 89). It was in 2010 that the OHR closed in Mostar, with the announcement made at Hotel Bristol on the bank of the Neretva, by High Representative Valentin Inzko (2010). However, this did not translate to a full withdrawal, and a small team of three was maintained in the city, housed by the Organisation for Security and Co-operation in Europe (OSCE) (Inzko

2010). Despite international and national efforts the political stagnation is maintained in Mostar. The ethnocratic maintenance of jurisdictional divisions, predominately due to the political narratives of the Hrvatska Demokratska Zajednica (HDZ-BiH—Croatian Democratic Union), and the Stranka Demokratske Akcije (SDA—Party of Democratic Action) has resulted in a political and city maintenance 'standstill' (Behram 2015). Furthermore, it is due to 'a lack of cross-party support' by the SDA and HDZ-BiH, that the city has not held a local election since 2008, and also did not participate in the most recent 2016 local elections[7] (OHR 2016). Moreover, in 2012, the City Statute was declared unconstitutional regarding 'a discrepancy in the number of votes required to elect councillors from each city area' (Inzko 2012). This figure 'varied from...7,000 to almost 30,000' while those who resided in the space of the 'central zone' could only elect councillors from the 'city-wide list' meaning they had no direct representation (Inzko 2012). Since 2012 the city council of Mostar has operated unmandated while also being ran by unelected and 'possibly illegal representatives' (Björkdahl and Gusic 2016: 89). Dissatisfaction with the ongoing political stagnation is evident, and in February 2014 protests spread across BiH. Starting in Tuzla, protests also occurred in Sarajevo and Mostar, with individuals setting fire to cantonal government offices (ICG 2014: 3). In Mostar, the 'offices of the leading Croat and Bosniak parties' were set on fire along with city administration offices (Fig. 4.2) (ICG 2014: 3).

The protests led to the establishment of 'plenums' which allowed for attendees to openly discuss issues. In Mostar the group met in the Spanish Square, a small group of mostly elderly residents still meet at the Spanish Square every Tuesday evening though numbers are significantly lower than they were initially (Spanish Square protestor 2015). In Mostar, the protests were considered by International Crisis Group to be a 'joint Croat-Bosniak affair' (ICG 2014: 4). However, despite what may be analysed as social unity, fostered through the economically motivated protests in 2014, institutional ethno-nationalistic divisions still dominate the system. As International Crisis Group reported in 2014 'the city statute cannot be amended until the (currently non-existent) city council formally adopts it' critically, this is something that Croat politicians 'promise never to do' (ICG 2014: 19). While politically the city remains divided, the 'governance framework' in place following OHR reform in 2004 means that the two sides are interactive (Björkdahl and Gusic 2016: 91). The institutional and social sub-divisions are evident

Fig. 4.2 City of Mostar Administration Offices (Photo taken by author in June 2014)

though, as '[t]he local ethno-nationalist elite seems to reject the internationally designed democratic governance system' while resistance to the ethnocracy is identifiable socially in Mostar and the across the country (Björkdahl and Gusic 2016: 91).

The milieu of responses to the governance system and those who seek to implement it is important to reflect on in drawing together this brief history of the staging of the city. An important point to take forward is the existence of the divided spatial narrative in the post-conflict establishment of the city, through institutional (top-down) actors, national, and international alike. Notably, the development of the city under various periods of colonial rule transformed the space over hundreds of years. The city was once emblematic of the ability of different social actors from religious and cultural backgrounds to live together, despite the tumultuous conflicts which waged in the city. Observably, there is a transgenerational transference of divisions from the 1992–1995

Bosnian war, and these divisions have shaped political discourse in the post-peace agreement Bosnia. This political deadlock facilitated by the Dayton Agreement has established conditions for the entrenchment of spatial and socio-political divisions. In Mostar this is materialised by electoral zones which correspond with the former divided municipalities, the divided education system, and divided public services. However, as illustrated by the centre Abrašević, the top-down staging of the city is, enacted against by some social actors who have sought to (and succeeded) in establishing shared space in the city, though such spaces may have a complex scripting.

NOTES

1. For example, Malcolm (2002: 13) summarises the medieval period of BiH as 'frequently confused and confusing'.
2. It should be noted that for a detailed account of the history of BiH you should consult Malcolm (2002).
3. Similarly, this tactic was deployed during the siege of Sarajevo; by the Army of the Serb Republic (VSP).
4. This term is discussed by Šavija-Valha (2017: 163–178) in its BCS form of 'raja' as a term used in contemporary times in (specifically) Sarajevo, which means 'common people' but which denotes commonality through embracing the self as a subject of irony and vetoing all pretences.
5. The current mayor, Ljubo Bešlić, has held office since 2004.
6. The bridge is also the location of the Mostar Diving Club, members of which, collect money from tourists before diving into the water. Due to the tradition of diving off the bridge in 2015, the Old Bridge has been one of the locations of the 'Red Bull Cliff Diving World Series' since 2015 (Red Bull 2018).
7. On the 2nd of November 2016 local youth staged an alternative election and invited residents to take part, this was held outside Mepas mall, in the Spanish Square, and online, called 'Izbori se za Mostar! (Elections are for Mostar)' (Alda 2016; Radio Sarajevo 2016).

BIBLIOGRAPHY

Alda. (2016). *Elect Mostar. Balkan Regional Platform for Youth Participation and Dialogue*. European Association for Local Democracy. Available from: http://www.alda-balkan-youth.eu/News/elect-mostar-97928. Accessed 15 Dec 2016.

Alibašić, A. (2014). Bosnia and Herzegovina. In J. Cesari (Ed.), *The Oxford Handbook of European Islam*. Oxford: Oxford University Press.

Annexe to the Dayton Agreement. (1995). *Dayton Agreement on Implementing the Federation of Bosnia and Herzegovina*. Dayton: Annex to the Dayton Agreement. Available from: http://www.peaceagreements.org/wview/2/Dayton+Agreement+on+Implementing+the+Federation+of+Bosnia+and+Herzegovina,+Dayton. Accessed 10 July 2015.

Annan, K. (2005, July 11). Secretary-General Statement on Srebrenica Anniversary Ceremony. SG/SM/9993. Available from: https://www.un.org/press/en/2005/sgsm9993.doc.htm. Accessed Jan 2018.

Architecture Week. (2004). *Stari Most—Mostar Reconnection*. Architecture Week No. 203. Available from: http://www.architectureweek.com/2004/0804/news_1-2.html. Accessed 20 Jan 2014.

Ashdown, P. (2004). *Statute of the City of Mostar*. Available from: https://www.mostar.ba/statut-181.html. Accessed Dec 2016.

Bar-Tal, D. (2013). *Intractable Conflicts: Socio-Psychological Foundations and Dynamics*. Cambridge: Cambridge University Press.

Behram, M. (2015, October 13). *Rats Infest Bosnia's Mostar Amid Political Deadlock*. Balkan Insight. Available from: http://www.balkaninsight.com/en/article/political-divisions-bring-health-hazards-to-southern-bosnian-city-10-12-2015. Accessed 20 Nov 2016.

Boyle, M. J. (2014). *Violence After War: Explaining Instability in Post-conflict States*. Baltimore: John Hopkins University Press.

Björkdahl, A. (2015). 'Two Schools Under One Roof' Unification in the Divided City of Mostar. In A. Björkdahl & L. Strömbom (Eds.), *Divided Cities, Governing Diversity*. Lund: Nordic Academic Press.

Björkdahl, A., & Gusic, I. (2016). Sites of Friction: Governance, Identity and Space in Mostar. In A. Björkdahl, K. Höglund, G. Millar, J. V. D. Lijn, & W. Verkoren (Eds.), *Peacebuilding and Friction: Global and Local Encounters in Post Conflict-Societies*. London: Routledge.

Bollens, S. A. (2007). *Cities, Nationalism and Democratisation*. London and New York: Routledge.

Bose, S. (2002). *Bosnia After Dayton: Nationalist Partition and International Intervention*. London: C. Hurst & Co.

Bosnia-Herzegovina Council of Ministers. (2004). Council of Europe, Report Submitted by Bosnia-Herzegovina Pursuant to Article 25, Paragraph 1 of the Framework Convention for the Protection of National Minorities. Available from: https://www.coe.int/t/dghl/monitoring/minorities/3_FCNMdocs/PDF_1st_SR_BiH_en.pdf. Accessed 20 Feb 2014.

Calame, J., & Charlesworth, E. (2011). *Divided Cities: Belfast, Beirut, Jerusalem, Mostar and Nicosia*: University of Pennsylvania Press.

Calame, J., & Pašić, A. (2009). *Post-conflict Reconstruction in Mostar: Cart Before the Horse* (Divided Cities/Contested States Working Paper. No. 7). Available from: http://www.conflictincities.org/PDFs/WorkingPaper7_26.3.09.pdf. Accessed 20 June 2014.

Carabelli, G. (2018). *The Divided City and the Grassroots: The (Un)making of Ethnic Divisions in Mostar.* Basingstoke: Palgrave Macmillan.

Chandler, D. (2000). *Bosnia: Faking Democracy After Dayton.* London and Sterling, VA: Pluto Press.

Chandler, D. (2006). *Peace Without Politics? Ten Years of International State-building in Bosnia.* Abington and New York: Routledge.

Djurasovic, A. (2016). *Ideology, Political Transitions and the City: The Case of Mostar, Bosnia and Herzegovina.* Abington, Oxon and New York: Routledge.

Domin, T. (2001a). *History of Bosnia and Herzegovina.* SFOR Informer #148 (Chapter 5). Available from: http://www.nato.int/sfor/indexinf/148/p04a/t02p04a.htm. Accessed 15 Apr 2014.

Domin, T. (2001b). *History of Bosnia and Herzegovina.* SFOR Informer #149 (Chapter 6). Available from: http://www.nato.int/sfor/indexinf/149/p04a/t02p04a.htm. Accessed 15 Apr 2014.

Drakulić, S. (1993, December 13). Falling Down: A Mostar Bridge Elegy. *The New Republic.* Available from: http://h-net.msu.edu/cgi-bin/logbrowse.pl?trx=vx&list=h-islamart&month=0811&week=b&msg=ltVciwbhxC-CxvN1//tgsag&user=&pw=. Accessed 20 June 2016.

EUAM. (1996). Special Report No 2/96 Concerning the Accounts of the Administrator and the European Union Administration, Mostar (EUAM) Accompanied by the Replies of the Commission and the Administrator of Mostar. *Official Journal C 287*, 30/09/1996 P. 0001–0021. Available from: http://eur-lex.europa.eu/legal-content/EN/TXT/?uri=CELEX%3A31996Y0930(01. Accessed 20 Aug 2016.

Forde, S. (2016). The Bridge on the Neretva: Stari Most as a Stage of Memory in Post-conflict Mostar, Bosnia-Herzegovina. *Cooperation and Conflict, 51*(4). Available from: http://journals.sagepub.com/doi/pdf/10.1177/0010836716652430. Accessed 16 Dec 2016.

Galtung, J. (1990, August). Cultural Violence. *Journal of Peace Research, 27*(3). Available from: http://www.jstor.org/stable/423472. Accessed June 2017.

Geohegan, P. (2014). Welcome to Brčko, Europe's only Free City and a Law Unto Itself. *The Guardian.* Available from: https://www.theguardian.com/cities/2014/may/14/brcko-bosnia-europe-only-free-city. Accessed Dec 2017.

Hill, C., & Smith, K. E. (2002). *European Foreign Policy: Key Documents.* London and New York: Routledge.

Hobsbawm, E. (2012). Introduction: Inventing Traditions. In E. Hobsbawm & T. Ranger (Eds.), *The Invention of Tradition.* Cambridge: Cambridge University Press.

Human Rights Watch. (1995). *The Fall of Srebrenica and the Failure of UN Peacekeeping, Bosnia Herzegovina* (Vol. 7, No. 13). Available from: https://www.hrw.org/sites/default/files/reports/bosnia1095web.pdf. Accessed 20 Nov 2016.

Implementation Force (IFOR). (1998). *The Rome Statement Reflecting the Work of the Joint Civilian Commission Sarajevo Compliance Conference.* Available from: http://www.nato.int/ifor/general/d960218d.htm. Accessed 28 Apr 2015.

International Crisis Group. (2000, April 19). *Reunifying Mostar: Opportunities for Progress* (Europe Report No. 90, ICG Balkans Report No. 90).

International Crisis Group. (2003, November 20). *Building Bridges in Mostar.* ICG Europe Report no. 150. Available from: http://www.crisisgroup.org/~/media/Files/europe/150_building_bridges_mostar.pdf. Accessed 20 June 2016.

International Crisis Group. (2014, July 10). *Bosnia's Future.* Europe Report No. 232. Available from: https://d2071andvip0wj.cloudfront.net/bosnia-s-future.pdf. Accessed 20 Sept 2016.

Internal Displacement Monitoring Centre. (2013, December 31). *Bosnia and Herzegovina: Internal displacement in Brief.* Available from: http://www.internal-displacement.org/europe-the-caucasus-and-central-asia/bosnia-and-herzegovina/summary. Accessed Dec 2017.

Inzko, V. (2010, June 10). *Speech by HR on the Closure of OHR Mostar.* Available from: http://www.ohr.int/?p=34433. Accessed 25 Sept 2016.

Inzko, V. (2012, April 12). *Mostar: Time for Compromise.* Available from: http://www.ohr.int/?p=32831. Accessed 20 Sept 2016.

Kappler, S. (2014). *Local Agency and Peacebuilding: EU and International Engagement in Bosnia-Herzegovina, Cyprus and South Africa. Re-Thinking Peace and Conflict Studies.* Basingstoke: Palgrave Macmillan.

Keil, S. & Perry, V. (2016). Introduction: State-Building and Democratisation in Bosnia and Herzegovina. In S. Kiel & V. Perry (Eds.), *State-Building and Democratisation in Bosnia and Herzegovina.* Surrey and Burlington: Routledge.

Kim, J. (2006). *Bosnia and the European Union Military Force (EUFOR): Post-NATO Peacekeeping.* Available from: https://www.fas.org/sgp/crs/row/RS21774.pdf. Accessed 21 Sept 2016.

Lovrenović, I. (2001). *Bosnia: A Cultural History.* New York: New York University Press.

Malcolm, N. (2002). *Bosnia: A Short History.* Kent: Papermac.

OHR. (2005, July 12). *OHR's Statement at the International Agencies' Joint Press Conference I Mostar.* Available from: http://www.ohr.int/?p=42287. Accessed 24 Sept. 2016.

OHR. (2016, October 21). 50th Report of the High Representative for Implementation of the Peace Agreement on Bosnia and Herzegovina to the

Secretary-General of the United Nations. Available from: http://www.ohr. int/?p=96473. Accessed 21 Nov 2016.

OKC Abrašević. (2016). *About Abrašević.* Available from: http://okcabrasevic. org/o-abrasevicu/. Accessed Nov 2016.

Palmberger, M. (2016). *How Generations Remember: Conflicting Histories and Shared Memories in Post-War Bosnia Herzegovina. Global Diversities.* London: Palgrave Macmillan.

Pamuk, Ş. (2000). *A Monetary History of the Ottoman Empire.* Cambridge: Cambridge University Press.

Participant K. (2015). *Interview in Mostar.*

Participant C. (2015). *Interview in Mostar.*

Pašić, A. (2004). *Conservation and Revitalisation of Historic Mostar.* Aga Khan Historic Cities Programme, p. 5. Available at: http://archnet.org/system/ publications/contents/3480/original/DPC1419.pdf?1384775278. Accessed 3 Jan 2014.

Perry, V. (2006). Democratic Ends, (Un)Democratic Means? Reflections on Democratisation in Brcko and Bosnia-Herzegovina. In M. A. Innes (Ed.), *Bosnia Security After Dayton: New Perspectives.* New York: Routledge.

Petrovic, J. (2012). *The Old Bridge of Mostar and Increasing Respect for Cultural Property in Armed Conflict.* Leiden: Martinus Nijhoff Publishers.

Radio Sarajevo. (2016). *Local Elections: Elections Are for Mostar: In the City on the Neretva Still Vote.* Available from: http://www.radiosarajevo.ba/vijesti/ bosna-i-hercegovina/izbori-se-za-mostar-u-gradu-na-neretvi-ipak-gla-saju/239891. Accessed 15 Dec 2016.

Ravn, B. (1997). *Bridge Over Troubled Waters.* SFOR Informer No. 11. May 28th. Available from: http://www.nato.int/sfor/engineers/mostarbridge/ introduction/introduc.htm. Accessed 20 Dec 2013.

Red Bull. (2018). *Cliff Diving World Series: Bosnia and Herzegovina.* Available from: https://www.redbull.com/us-en/events/red-bull-cliff-diving-bosnia. Accessed Apr 2018.

Riedlmayer, A. (2002). From the Ashes: The Past and Future of Bosnia's Cultural Heritage. In M. Shatzmiller (Ed.), *Islam and Bosnia: Conflict Resolution and Foreign Policy in Multi-Ethnic States.* Montreal, Quebec: McGill-Queen's University Press.

Said, E. W. (1979). *Orientalism.* New York: Random House.

Said, E. W (2000). Invention, Memory, and Place. *Critical Inquiry, 26*(2). Available from: https://www.jstor.org/stable/1344120?seq=1#page_scan_ tab_contents. Accessed 20 June 2015.

Šavija-Valha, N. (2017). Raja: The Ironic Subject of Everyday Life in Sarajevo. In S. Jansen, Č. Brković, & V. Čelebičić (Eds.), *Negotiating Social Relations in Bosnia-Herzegovina, Semi-peripheral Entanglements.* London and New York: Routledge.

Selimovic, J. M. (2016). Frictional Commemoration: Local Agency and Cosmopolitan Politics at Memorial Sites in Bosnia-Herzegovina and Rwanda. In A. Björkdahl, K. Höglund, G. Millar, J. V. D. Lijn, & W. Verkoren (Eds.), *Peacebuilding and Friction: Global and Local Encounters in Post Conflict— Societies*. London: Routledge.

Siani-Davies, P. (2004). *International Intervention in the Balkans Since 1995*. London and New York: Routledge.

Spanish Square protestor (2015). *Informal Conversation*.

Stone, P. G., & Bajjaly, J. F. (2008). *The Destruction of Cultural Heritage in Iraq*. Woodbridge: Boydell & Brewer.

Todorova, M. (1997). *Imagining the Balkans*. Oxford and New York: Oxford University Press.

Todorova, M. (2009). *Imagining the Balkans*. Oxford and New York: Oxford University Press.

UN. (1995). *General Framework Agreement for Peace in Bosnia-Herzegovina*. November 30th. Available from: http://peacemaker.un.org/sites/peace-maker.un.org/files/BA_951121_DaytonAgreement.pdf. Accessed July 2015.

UNESCO. (2005, July 15). *The Old Bridge Area of the City of Mostar*. World Heritage Scanned Nomination. Available from: http://whc.unesco.org/uploads/nominations/946rev.pdf. Accessed 14 June 2014.

UNESCO. (2013). *Introducing UNESCO*. Available from: http://en.unesco.org/about-us/introducing-unesco. Accessed 2 Jan 2014.

UNESCO. (2016). *Tangible Cultural Heritage*. Available from: http://www.unesco.org/new/en/cairo/culture/tangible-cultural-heritage/. Accessed Dec 2016.

Velikonja, M. (2003). *Religious Separation and Political Intolerance in Bosnia-Herzegovina*. College Station: Texas A&M University Press.

Vetters, L. (2013). The Power of Administrative Categories: Emerging Notions of Citizenship in the Divided City of Mostar. In S. Stroschein (Ed.), *Governance in Ethnically Mixed Cities*. Cornwall: Routledge.

Willigen, N. V. (2013). *Peacebuilding and International Administration: The Cases of Bosnia and Herzegovina and Kosovo*. New York: Routledge.

Winterstein, N. (2003). *Commission for Reforming the City of Mostar* (Recommendations of the Commission Report of the Chairman). Available from: http://www.ohr.int/archive/report-mostar/pdf/Reforming%20Mostar-Report%20(EN).pdf. Accessed 20 June 2015.

Yarwood, J. R., Seebacher, A., Strufe, N., & Wolfram, H. (1999). *Rebuilding Mostar: Urban Reconstruction in a War Zone*. Liverpool: Liverpool University Press.

Živanović, M. (2015). Discrimination: From Construction to Deconstruction an Essay on the Prospects of Reconciliation in Bosnia-Herzegovina 20 years after Dayton. In M. Fisher & O. Simic (Eds.), *Transitional Justice and Reconciliation: Lessons from the Balkans*. New York: Routledge.

CHAPTER 5

Spatial Narratives in the City: *Art, Graffiti, and Movement*

'Every story is a travel story- a spatial practice' (De Certeau 1984: 115). Cities become identified as such spatial forms due to the resources, structures, and crucially, the population within a defined space. But much like the principle of space-time relativity, two actors looking at one space, moving on slightly different paths will view the space in differing ways. We are all moving along our own spatial trajectories and have all arrived in one particular space or another, having followed unique courses. In short, this is our positionality, which is a spatially situated term that reflects how our 'ascribed' identities (race, gender, sexuality, class) interact with others (Newman 2014: 24). In many ways positionality affects what is visible to us, and for research this has implications regarding how our position, relative to others, impacts on what we identify as researchers (but also how others view us). Therefore, untangling perceptions and narratives of a city becomes a complex task, which demonstrates the importance of looking deeper at Mostar, beyond the narratives of the divided city. That said, it is impossible to tell the story of a city in a single chapter, or a book, as there are only so many spatial narratives which can be incorporated.[1]

As previously outlined, the ethno-nationalistic divisions in the city of Mostar can be considered as enduring due to a combination of issues stemming from political instrumentalism, physicalised through the maintenance of a spatially realised discourse of division. Divisions manifest in all cities because they are spaces of diversity, there are many well-known divided cities such as Jerusalem, Cape Town, Mostar, Belfast and

© The Author(s) 2019   105
S. Forde, *Movement as Conflict Transformation*, Rethinking Peace and Conflict Studies, https://doi.org/10.1007/978-3-319-92660-5_5

Nicosia. However, most cities can be considered divided cities, they have just not suffered a physical intrastate conflict, for example, London or San Francisco, though there may be no walls, or conflict divisions, these cities have distinct spatial boundaries of inequality. Space, movement, and use of space, is inherently divisive however some divisions such as gendered divisions and economic divisions are naturalised in most cities and spaces. The categorisation of a divided city relegates a space as problematically and intractably divided, this definition also overlooks the agency of individuals in such cities and assumes that those in post-conflict space are purely passive. However, this is a generalisation and to some extent a process of 'othering' (Said 1979). This overlooks systemic inequalities and pervasive power structures, but also the agentive capabilities of individuals to subvert top-down power structures in ways that, due to traditional routes of research engaging with such power structures, researchers previously have discounted. It is in the everyday that peace, positive or negative (Galtung 1967: 12) is lived, this is not to say that institutional and educational divisions do not impact on the everyday but to observe that individuals have the agency to traverse some divisions. Through movement to shared spaces, and in the establishment of shared spaces, social actors have rescripted divisive space in the city of Mostar. When interviewed, the participants' social movement was generally without concern of the institutional divide. However, as noted by some participants, their movement had previously not crossed the institutional divide of the city. From this it can be briefly summarised that involvement in shared spaces, transformed divisional movement. Of course, the very movement to the space itself may not be transformative alone, but the social interactions and experiences in that space can impact on wider spatial movement in divided spaces. The importance of memory and narratives of experiences in space is demonstrated through the transgenerational impact of the ethnic conflict which affects present day movement. While many participants did not directly experience the conflict first hand in the city, it is indicative of the transgenerational permeation of the divisions that these participants have previously experienced a division in their social movement which correlates with the ethno-nationalist divide. Research participants were conscious of the divisions in the city, though all moved physically within shared spaces which aim to traverse the ethnic divisions. In moving through and within these spaces, individuals oscillated between movements which are non-divisional, while working and/or living through spatial practice to transgress divisions which persist in the city. Through this, all participants showed an awareness of the divided city status and the purpose of the research project taking place in Mostar.

Physical, Temporal and Conceptual Spaces

In spatial analysis, a distinction can be made between the physical and the figurative, between the actual tangible space (the staging) and the use of that space through meaning ascribed to it (the scripting)—the latter sometimes materialises, though not always. There are numerous spaces which can be analysed as shared spaces in the city of Mostar. Physically, the locations of OKC Abrašević, the United World College (UWC) in Mostar (an international college), the Pavarotti Music School and the Mostar Rock School exist as shared spaces. These spaces focus on youth and are cooperative and interactive in the city, along with the unfixed temporally operational Street Arts Festival (SAF). Together these spaces exist as an informal peace network in that they are collaborative but also independently support the 'objectives' of each space. In particular, OKC Abrašević is discussed as a hub which facilitates the creation of shared spaces such as the SAF. Additionally, the Rock School and UWC have both been involved in the SAF, with young adults who attend the educational institutions performing music or producing pieces of art during the festival. The engagement of UWC students in rescripting public space is indicative of the founding purpose of the UWC, as an educational facility which interacts with post-conflict reconciliation. Furthermore, the focus of SAF on youth engagement is important as youth unemployment is significantly high in BiH, as of 2017 the modelled International Labour Organisation estimate of youth unemployment (ages 15–24) was 67.5% (World Bank 2017). This is an ongoing problem in BiH, and poses multiple social problems including a lack of economic engagement and potential disenchantment with society (World Bank 2017).

Therefore, the low cost or free access space that this network provides for cross divide interaction is vitally important in the city. Though, as with all spaces, the scripting of each location translates to duel accessibility and inaccessibility of space. For example, as previously noted, the Abrašević youth centre exists as a location open to all in the city and is an important space for facilitating shared experiences of space. However, regardless of spatial purpose, no public space exists without social boundaries and as a result, even a space intended for all is a closed space for some, as some individuals may feel "out of place". This is not to negate the important work that the centre does, but to note that all spaces are staged, and in this case, scripted in a particular social context which creates a liminal boundary. However, there can exist multiple spaces in one place, as spaces are not a fixed building or an area but are constantly rescripted, dependent on the social actors who use the space, the experiences of the space and the meaning, and memory derived from this.

STAGING AND SCRIPTING THE CITY

The construction of public space tells the story of a spatiality through memorials and statues, buildings and roads, which establishes heritage and anchors individuals to the space. The staging of public space reflects the institutional narrative and aims to direct the social usage of the space. As space and identity are interdependent, the staging attempts to control the use of space, but also the identity of those who use that space. In the context of a city, the staging of the city as divided reinforces the link between political goals and physical space, to the formation of identity. While institutional staging of space is a particular issue in divided cities, the social use of space does not always correlate as every city has a milieu of spatial narratives. This is true of all cities and these narratives are sometimes visualised, a common form of this is graffiti (and sometimes stickers and symbols) and street art. In Mostar, there is both graffiti and street art, the latter is artwork completed during the SAF which runs each year in the city. While both types of urban engagement could be described as graffiti, in this work graffiti is considered informal 'tagging' and street art is attributed to work that is produced during the festival. The art produced during SAF can be considered somewhat neutral, or critical of the socio-political division, this is reflective of the mission of the SAF. Whereas, graffiti in the city is frequently ethnonationalistically divisional or fascist, and is sometimes countered by anti-fascist graffiti. While anti-fascist graffiti has a different narrative of the use of space, in that it opposes fascist divisional threats of violence, due to its informality it will be looked at alongside the divisive graffiti. Cities are full of informal communication networks such as stickers, posters, street art, and graffiti. In cities, such symbols indicate, others that think like us—or unlike us, use this space too. As Tonkiss (2005: 140) observes, city space is a canvas through which 'demands might be advanced, identities inscribed, and challenges issued'. This works both ways in Mostar, the street art challenges the divisive nature of the city space, while divisive graffiti seeks to establish spatial norms about the ownership of space. As the division in Mostar is not physicalised by walls, informal signs of division attempt to establish the use of space which sometimes correlates with the institutional divisions in the city. In this capacity, individuals can rescript space to correlate with the institutional staging of space and this is not limited to post-conflict spaces. Fundamentally, the street art, which does not represent ethno-nationalistic divisions, is produced during the SAF and has been selected by organisers. To participate in the festival, artists apply to contribute their work to the city, and organisers facilitate this by finding a suitable place for each artwork in the city. While the divisional graffiti, stickers and symbols, and the street art, all represent a social scripting of the physical space of the city, the narratives are somewhat divergent. These visual signifiers present the fact that social scripting can correlate with institutional divisions or can transgress such divisions. The extent to which these markers influence movement is socially space-time dependent, in that it depends very much on the observer and how they process the scripting.

Rescripting Space and Art

The SAF in Mostar was established by several OKC Abrašević members including participant G and has increased annually in size and scope. Every year the festival has an open call for international, national and local artists to participate, including not only painted street art (Fig. 5.1), but street performances, workshops, music and other activities. SAF was established to promote the use of public space by youth in the city, and students and local youth are encouraged to interact with the festival either through participation or volunteering (G Interview 2 2016). The festival has unique access to the city space, as due to local government approval, each year the whole of the city becomes a canvas for the festival, subject to private landowners' approval of the use of space (G Interview 2016). As performances and workshops are also organised during the festival, which is temporal in nature, it can be considered to be a shifting space. The festival is socially supported by volunteers and is part of a network of spaces working to overcome divisions, supported by local organisations such as the UWC Mostar and OKC Abrašević. The "Street Art collection" (Fig. 5.2) is constantly growing around the city, according to G the process of placing the art is typically guided by

Fig. 5.1 'I heart Mostar' by Bibbito (Photo taken by author in September 2015)

'intuition' and starts with G walking through the city to 'see where art can fit' (Interview 2 2016). Following this, permission is sought from whoever owns the building/the space (G Interview 2 2016). As a result of the visibility of the art, the SAF can be considered a key "actor" of social rescripting in the city and one that is capable of influencing the use of the public space in Mostar. The festival facilitates a temporally dependent engagement with the city space. Fundamentally, street art and graffiti can be considered visualisation of the concept of rescripting as they both present a social narrative of public space. In the case of the street art, the work provides a reconceptualisation of the use of public space in the

Fig. 5.2 Staklena Banka (Glass Bank collection) (Photo taken by author in September 2015)

city of Mostar, beyond the staging of space. Fundamentally, the social experience of movement during the festival, and the art that remains, holds potential for the rescripting of meaning in space, through movement inspired by art. Street art is a visible marker of social movement in space and can be powerful in signalling a commonality in the definition of space. Critically, the festival works as a semi-permanent space which engages "both sides" of the city through the use of the whole city space, for participation in the festival.

ART AND THE OLD GLASS BANK (STAKLENA BANKA)

The Old Glass Bank, named for its pre-war glazed façade, is situated on one of the front lines of the conflict. It is a popular space in the city and it is located near a popular meeting and transport location near the Spanish Square. During the 1992–1995 war, snipers utilised the building due to the vantage point it offered across the city. The building has been abandoned since the war and due to potentially unexploded munitions, and the insecure structure, as a safety precaution it is advised that people do not enter the building though as explored further in Chapter 6 this advice is not always followed. The volume of art work at the Old Glass Bank is representative of the visibility and therefore, importance of the space in the city. The Old Glass Bank is a central space and the derelict building provides a blank canvas for artists. The building functions as an open-air exhibition of art, and a key location for street art in the city of Mostar. There are numerous pieces of street art visible from the outside of the bank which rescripts the institutional staging of the space as derelict and transforms the appearance of the building (Figs. 5.3, 5.4, 5.5).

As can be seen in Fig. (5.3) the street art with a positive message exists in close proximity to graffiti, in this example, sectarian graffiti. The use of IRA graffiti or 'Irish Republican Army' parallels the religious

Fig. 5.3 'Pick your glasses' by Mišela Boras (Photo taken by author in June 2014)

divide in the Northern Irish Conflict—with the divide in the Bosnian war. 'We are all living under the same sky', (Fig. 5.5) is a piece by Ale Senso, an Italian street artist. The piece is written in three languages BSC, German and English, pictured above is the English and BSC version. It is located in the old entrance to the Old Glass Bank, the three remaining pillars of the entrance, provides the canvas for the three-language format of the art. The piece evokes a message of shared humanity, in the spatial context of the derelict building and the continued divisions in the city. The message of the piece rescripts the space, through providing a message of peace in a space still staged as derelict by the conflict. Notably, the art has the potential to be repurposed by individuals who view, and (as I have done) photograph the space. It is important to note that the directed capturing of an image conveys the narrative of the photographer (Dauphinee 2007: 40). The photographs fundamentally provide a snap shot of the space and are directed by myself, as Sontag

Fig. 5.4 'Hidden Ambitions' by Gigo Propaganda (Photo taken by author in June 2014)

Fig. 5.5 'We are all living under the same sky' by Ale Senso (Photo taken by author in June 2014)

(1979: 6) observes 'photographs are as much an interpretation of the world as paintings and drawings are'. This is more so true in contemporary times, with electronic photographs, for example, Fig. 5.3 could have been cropped to exclude the IRA graffiti. It should be noted that it is not the same picture used for the cover of the book and the IRA graffiti may or may not have been in shot when that photo was taken.

However, while taking into account the ambiguity of the reception of street art, this is not to say that the art does or does not facilitate conflict transformation, but it does change the narrative of the space, which dependent on the viewer, then has the potential to influence perceptions of space. The perception of the space will not be homogenous, but it is important as it represents a social narrative visible to other users of the space. Fundamentally, the street art, in the format of written English becomes an internationally accessible piece of 'knowledge' about the spatiality. As the location of the art is locally led, it can be considered to be reflective of the perceptions of the SAF team. Independent of intentionality or reception, the art provides a narrative of shared humanity, which rescripts the space staged as derelict from the conflict and repurposes the space from a derelict building, to a space for art.

'Hidden Ambitions' (Fig. 5.4) is part of a piece by artist Gigo Propaganda a Mostar-born resident of Germany (Waende Süedost 2011). These words accompany artistic figures which are not visible from the outside of the bank. The phrase 'Hidden Ambitions' is the title of the artistic figures which have also been produced by the artist in Germany, which Gigo Propaganda describes as '[s]cary looking, vicious and incomprehensible beings [which] occupy both the picture and life' (Waende Süedost 2011). The artist was born in in Mostar in 1979, in an interview for Waende Süedost, Gigo discusses his movement from the city;

[i]n 1991, the year the communist system of my native country collapsed, I moved with my family to Essen, Germany...There I encountered new and different social and political ideas and an unknown multicultural society. (Waende Süedost 2011)

Gigo considers the focus of his artwork to be 'issues of rapid changing meanings of places and...the alienation of citizens from their environment' (Waende Süedost 2011). Notably the words 'Hidden Ambitions' are visible from the outside of the building but the self-described 'scary' figures that make up the rest of the work, are not (Waende Süedost 2011). As such, in the context of the Old Glass Bank, the meaning of the words is up to the interpretation of the observer and contextualised by their interpretation of the space, and their own spatial trajectory. Fundamentally, there are unfulfilled ambitions of the space this is reflected by participants who considered the potential of the building, in particular J and H, who noted that previously the use of the building was explored previously as a site for a youth cultural centre (Interview 2015). Similarly, participant G proposed that the space of the Old Glass Bank had the potential to become a central space for youth (Interview 2015).

The transformative effect of the art can be compared to Bell's (2009: 141) discussion of artistic interventions in the city of Sarajevo (as part of the city's Winter Festival or Sarajevska Zima) as a 're-appropriation of public spaces'. During the festival in Sarajevo, public space and streets are transformed into 'galleries of inclusion' (Bell 2009: 141). In further comparison to the SAF Mostar, the Sarajevo Winter Festival included international artists, which Bell (2009: 136) focuses on, noting that 'the foreign artists' re-appropriation of public spaces demonstrates their own compassionate responses to the war experiences of Bosnia-Herzegovina'. In many ways this can be compared to the SAF, as some artists are international, including those who take part via the UWC. It can also be noted that though some of the narratives of the pieces of work are scripted by international artists, the local team directs what art is included and where it goes. Furthermore, from an anti-essentialist standpoint, space is inherently global and local at the same time and the two are inseparably intertwined (Massey 1991). There is however a 'time-space compression', which is where a tension may be identified in the free-flowing movement of international actors, be they artists or researchers to divided cities, and most crucially from divided cities as 'control over mobility both reflects and reinforces power' (Massey 1991: 25). In focus, mobility for some is always in juxtaposition to immobility for others, and the two are sometimes interrelated;

> Some initiate flows and movement, others don't; some are more on the receiving-end of it than others; some are effectively imprisoned by it. (Massey 1991: 25)

The importance of space and movement with regard to the festival can be seen physically at a local level and also internationally. The festival draws international artists, and transforms the city centre, changing the narrative of the space, which is important for tourism in the city. As Naef and Ploner observed in 2016 (2016: 181). BiH only recently achieved 'its pre-war tourism market a few years ago'. The former-Yugoslavia region is popular with young travellers, in particular backpackers, there is a notable 'bohemian appriciation' of the Balkans among 'international artists, film makers [...] photographers' and others, this can be considered to extend also to tourism (Naef and Ploner 2016: 185). However, Volčič (2007: 29) observes the trend of the commercialisation of the former Yugoslavia, with one tourist agency in Ljubljana advertising 'what it describes as a "real Yugoslav experience."' This Yugo-nostalgia, extends far beyond its long erased borders and there is an affinity for some travellers with the cosmopolitan ideals of sub-cultures within the early 1990s. In focus, 'Rock 'n' roll musicians[…]mainly opposed extreme nationalistic tendancies and resisted the Milošević regime' (Kaneva 2011: 215). This emerged from a historical period of SFRY in which the state attmpted to popularise folk music, 'traditional rural music' was largely replaced by 'newly composed folk music' which became a vehicle for the nationalist agenda which preceded the 1992–1995 war (Kaneva 2011: 214). The culture war which was ongoing during the Cold War resulted in the Yugoslavian music being influenced by Western music. Notably the rise in popularity of Rock 'n' Roll is observed by Mišna (2016: 83, 80) as one that legimised the importance of youth culture and, in particular, its '*conscience collective*' which was characterised by 'anti-establishment [rebellion] out of genuine (ethical) impulses and a sense of commitment to those'. The messages of the art included in the festival could be described as artistic protests and anti-establishment in that they create an alternative narrative to the division in the city. The art and space are interactive, the space does not complete or define its meaning, but the art contributes a script to its spatial location. As highlighted above, the SAF has a distinct international dimension due to participants and some of the language used in the artwork. The use of German and English makes the art accessible to international audiences and international individuals using the city space. As such, it is liminal in reaching across spatialities, while being physically fixed in BiH.

ART AND THE OLD LIBRARY

The Old Library is an Austro-Hungarian building in Mostar, which as the name suggests, was once a library but is now derelict. The space was introduced to the research project as a potential space for renovation during the third field research visit in September 2015. Earlier that year in the summer, the 2015 SAF used the Old Library as a location for a workshop, which brought movement to the derelict building. However, the space has been used previously, as J recalled, 'I love these places that serve even in ruins, I think we used the library first in 2000, later as a location for the SAF' (Interview 2 2015). As shown in Figs. 5.6, 5.7, 5.8, and 5.9, the use of the space as a location in the SAF, transformed the Old Library, this included the marking of the space as a 'fascism free zone' (Fig. 5.9).

In reflecting on the increased focus of the Old Library as a potential social space in the city, the inclusion of the building in the SAF increased movement to, and therefore, spatial awareness of the unfulfilled potential of space due to its current condition. Consequently, numerous

Fig. 5.6 Pre-2015 workshop pictures of the Old Library (Photo taken by author in June 2014)

Fig. 5.7 Pre-2015 workshop pictures of the Old Library (Photo taken by author in June 2014)

participants outlined the potential of the space as one that may be rescripted in its former function or a new function. Though few discussed the potential of the building in-depth, those who did, discussed issues with potential transformation. In particular, according to participant J there are issues regarding the ownership of the space;

Fig. 5.8 The Old Library September 2015 Street Arts Festival (Photo by the author in September 2015)

> There are some property issues with the Old Library, there were plans to turn it into an international primary school, renovation [will take] 5–6 or 10 years. God knows when this will be planned. (Interview 2 2015)

The issue of ownership was also discussed by participant L (Interview 2015) as representative of a long-standing problem with responsibility and ownership of property in the city. As a result, L considered that if owners neglect investment in property upkeep, the council should 'intervene to get rid of them' in order to overcome the current stalemate of property owners neglecting derelict spaces (Interview 2015). This reflects a city-wide problem in Mostar with regard to investment, and the transparency of the ownership of properties. The latter issue with regard to the ownership of space, and particularly space which is derelict was also observed as a wider city issue by participant R;

Fig. 5.9 'Fascism free zone' on the steps of the Old Library (Photo taken by author in September 2015)

> I think today in the City of Mostar, urban planning and development does not exist in the true sense because plans are adjusted to investor's wishes and [the] investor is always related to politicians and they do not respect the interests of the citizens, rather [...] only their private interests. (Email Correspondence 2015)

However, notably this is not a condition exclusive to the city of Mostar but occurs 'in all cities in ex-Yugoslavia countries [...] none of the officials is [sic] going to explain it so, but it is true' (R Email Correspondence 2015). In Mostar there is still a significant number of buildings damaged by the war, which maintains the narrative of the conflict. This issue of a lack of investment and renovation coupled with stalled construction projects in the city was highlighted by numerous participants. The lack of spatial reconstruction has an impact not only on the social use of the city space but on generations growing up, as participant L observed 'the real problem is a generation born with the ruins'

(Interview 2015). While L applies this to buildings and urban structures, it can also be readily applied to the political maintenance of divisions in BiH, which function similarly to ruins through maintaining the divisions of the conflict and dissuading social participation. Notably, though issues with ownership, institutional issues with planning, and political interests traverse spaces, the social response is unique to each locale. In this, it can be put forward that the interactivity with the abandoned, disused and neglected buildings is unique to Mostar, as the spaces and actors are unique to Mostar. Comparative to the Old Glass Bank, the Old Library is also staged as a space of non-social movement, however, through social use and rescripting of the spaces as part of SAF, both have been used in an alternative function to that which they are staged. Through this they provided, and provide, opportunities for social actors to traverse the ethno-national division and other sub-divisions.

SIGNS, SYMBOLS AND STICKERS OF A DIVIDED MOSTAR

While there is a great deal of art in the city which visualises the conflict transformation taking place, there are also signs and symbols of division. Much like street art in the city, stickers and graffiti, signs, and murals create a narrative of the space which can correlate or clash with institutional ethno-nationalistic divisions. A main source of graffiti stems from extreme football fans, Ultras. There are two local teams in Mostar, Fudbalski Klub Velež (FK Velež Mostar (from the Bosniak side)) and Hrvatski Športski Klub Zrinjski (HŠK Zrinjski Mostar (from the Croat side)). Additionally, there are two Ultras groups in the city aligned with each football team, Red Army Mostar (associated with Velež and the East) and Ultras (associated with Zrinjski and the West) both have marked the East and West sides of the city respectively, with graffiti.[2] Notable examples of divisive signifiers (Fig. 5.10 and Fig. 5.11) include stickers such as 'Ustaše Mostar HATE ANTIFA'[3] and 'Zapadni [West] Mostar' appearing alongside "football club" stickers such as 'Zrinjski Mostar.' While Ultras who support Zrinjski, through stickers and murals align themselves with right wing politics (for example, the association with Ustaše and a skin head mural in the West of the city), the Red Army Ultras align themselves with more leftist politics visualized through the usage of the term 'urban guerilla' (Fig. 5.12) and caricatures of Che Guevara (Wabl 2017). In a discussion about the graffiti, J noted the frequent visibility of graffiti in the city, '[t]hey are identifiable

Fig. 5.10 Croat Ultras stickers visible on road signs in the West of the city (Photo taken by author in April 2015)

all over the city, these symbols can also appear to be benign (like football clubs) but in reality, not so much' (Interview 2 2015). Fundamentally, the marking of space as distinct through graffiti, stickers, and signs, indicates ownership of the space and demonstrates a social scripting, which aims to establish a division of space. This is not limited to Mostar, it can also be seen in other divided cities such as Belfast, through the visibility

Fig. 5.11 Croat Ultras stickers visible on road signs in the West of the city (Photo taken by author in April 2015)

of the Irish tricolour and the Union Flag of the UK, and the respected colours of these flags on curbs, lampposts and walls. There is generally no individual authorship of graffiti (not including personalised tagging) or stickers, instead they speak to a broader group identity that may be ideological in nature. This is not always divisional or reflective of the ethno-national divisions, for example, in Mostar anti-fascists are also

Fig. 5.12 1981 (Velež Mostar supporters 'Red Army' founding year) 'Urban Guerrilla graffiti' in the East of the city (Photo taken by author in June 2014)

visible through their use of graffiti and stencils. Therefore, in the city, these informal spatial indicators of focus, are scripted by three broad, but not comprehensive, categories (the first two identifiably overlap in some cases). These categories are, Ultras (football fans who frequently instigate violence), individuals who hold fascist ideologies, and those who are part of the anti-fascist movement. Fundamentally, the presence of the Ultras (which the graffiti narrates) can impact on the movement of others or their use of space;

> I don't mind going different places but if Velež and Zrinjski play…they don't usually fight but hooligans get beat a lot. The Ultras gather like 10–20m away. I know nothing is going to happen, but I don't feel comfortable. If a bigger club plays in the stadium sometimes they burn things. (Interview with D and E 2015)

As demonstrated through D's narrative sometimes football matches
end in violence. It is worth noting that the Ultras who typically instigate violence and can be considered distinct from other fans, not only
in their behaviour, but that the space that they occupy (Interview D
and E 2015). Though the Ultras occupy a space distinct from other
fans, it is noticeable that the proximity of the group still creates a tension around the possibility of violence occurring. Before the 1992–
1995 war there was only one football team in the city, FK Velež, which
was a popular team in the city and the wider Yugoslavia, its red strip
symbolic of the clubs' orientation towards 'leftist politics and multi-ethnic ideals' (Wabl 2017). This is still representative of the clubs
today as they 'have also become symbols of left and right ideologies'
(J Interview 2 2015). The Ultras in Mostar (from both teams) are
particularly visible through stickers and graffiti which correlate with
the 1992–1995 conflict divisions in the city. Figures 5.10, 5.11, 5.12,
5.13, 5.14, and 5.15, are some examples of graffiti that characterise the ideological divisions in the city. The divisional signifiers exist as
a social scripting of the space which correlates with institutional divisions, such scripting aims to define the historical heritage of the space,
and significantly, who is expected to use the space. In 2016, xenopho-
bic posters were put up in the city which depicted a group of women
wearing niqabs, with the universal symbol for no (red circle with a line
through it) over the top and the profanity, 'Fuck off' above the picture
(Kajan 2016). The offensive, and in this context, deliberately threatening phrase, is spatially orientated, in that it is used socially to tell someone (or in this case a group of people) to leave the space in which the
individual communicating the message exists in. Such spatialisations of
xenophobia demonstrates the prevalence of the conflict narrative and
explicitly spatialises the threat of violence. The prevalence of fascism
in the city can be traced to WWII and the conflict which occurred in
Mostar. The influence of the Nazi puppet state, the Independent State
of Croatia or Nezavisna Država Hrvatska (NDH), is evident through
the visibility of fascist graffiti and also streets which are named after
members of the NDH (Kajan 2016). Fascist graffiti is also visible at
the Partisan memorial (discussed further in Chapter 6) in the West of
the city which was constructed for the Partisans who defended Mostar
during the Second World War. Despite attempts to keep the memorial graffiti free, fascist groups have repeatedly established their visibility through graffitied swastikas and other fascist symbols alongside

Fig. 5.13 '1994 Ultras Centar 2' mural in the West of the city (Photo taken by author in September 2015)

phrases such as 'Skinheads Mostar' (Kajan 2016). In discussing the graffiti around the Partisan memorial, participant J noted that the fascists and antifascists are interactive in the city through 'a fight of symbols. Right to left' (J Interview 1 2015). While graffiti, stickers, and symbols of division are common place in cities as they are heterogenous spaces, the presence of fascism demonstrates xenophobia in the city which is overlaid with the divisions from the 1992–1995 conflict. The presence of such signs, and symbols, tells a divisive narrative of the city which through scripting attempts to replicate the political boundary of ethno-nationalistic divisions. The examples of graffiti and stickers, some of which have been demonstrated in Figs. 5.10–5.15, are visible all over the city, with a general East/West divide which marks out the territory of the respective Ultras and Antifa. What is notable is that there is an ongoing conflict of symbols occurring in the city, through posters, stickers, and graffiti which establish spatial ownership, sometimes through

Fig. 5.14 Antifa spray sign in the West of the city (Photo taken by author in June 2014)

intimidation and the threat of violence. This is can be juxtaposed with the SAF which facilitates the production of art in the city as a spatial strategy to increase movement and use of the city space. Critically, divisional signs and symbols indicate the social scripting of space, in some examples the scripting reaffirms institutional divides, while street art (through the festival) transforms the scripting of the space through

Fig. 5.15 'Smash Fascism'. Former front line of the conflict in the West of the city (Photo taken by author in June 2014)

the collaborative nature of the production of the work and in doing so encourages movement and transforms space.

RECONSTRUCTION AND CORRUPTION IN THE CITY

In Mostar, a key issue regarding the use of space is the divided political system, the presence of corruption maintains the system, which deadlocks the political system but also reconstruction. In outlining the use of the city space, the interrelation of the issues of the use of the land, the ecology of the city and the wider country are apparent. Frequently, the institutional divides in the city are visualised through disputes regarding the responsibility of who collects what waste, and in what city area, which results in an accumulation of waste (Hopkins 2015) this was mentioned by participant H who noted 'there is trash everywhere, it is a big ecological crisis waiting [sic]' (Interview 2015). The general neglect of the local area is evident, alongside stalled reconstruction projects, while this denigration of public space is exacerbated by the privatisation of space and a lack of regulation in the post-war, post-socialist era;

J: In the time of socialism you were not allowed to build where and when you wanted. People built where they could it was not compensated but there was advice [now] most permission for the building is from political/economic connections, there is no planning.

I: Yes this is what I know also.

J: You used to need a permit [...]

I: New buildings are emerging all the time they are just cramming an apartment building anywhere.

J: The city used to be very unified, red roofs, white buildings, now there are lollipop facades. Citizens were responsible before but now they have a goldfish memory. (Workshop 2015)

In the above, there is a notably nostalgia for the visual commonality of neighbourhoods during the Yugoslav era. As noted by J housing was not always subsidised in the Yugoslav era (Archer 2018: 142). However unregulated housing projects did occur during this time through self-constructed housing which were typically 'neither entirely legal nor outright illegal but rather existed on a continuum', construction of such housing was frequently socially cooperative and involved a skills exchange (Archer 2018: 144). The post-war narrative of the construction process as one that stems from 'political/economic connections' references corruption in the political system as permeating the staging of public space (J Workshop 2015). Those in power directly impact on the use and definition of space, and this is considered an unrepresentative process as 'the city council elects the mayor who is not accountable to us but to the council and serves who is in power [...] the mayor was re-elected though there was no way for us to participate in the election in 2013' (J Interview 2015). The entrenchment of corruption was also discussed by participant M who surmised, '[t]here is corruption at all levels, the human rights, economic corruption, down low corruption and top-down corruption' (Interview 2015). Corruption has as a pervasive impact on post-conflict reconstruction and spatial transformation in the city, this was further discussed by J as linked to the conflict through war profiteers;

The current system needs starting capital the majority who have [starting capital] are war profiteers, they sold property evaded paying taxes, stealing the earth. People who have starting capital in and after the war, are experiencing a nice life, what we should all be experiencing. (Interview 2 2015)

The current use of the public space is directed by the previous staging of the public space (through the war) and according to J, the actors who performed certain roles benefitted financially from the conflict and benefit further in the post-conflict environment. Fundamentally, J's reflection indicates a feeling of a lack of "justice" post-conflict as she considered that the privileged position of war profiteers (Interview 2 2015). The occurrence of war profiteers in Bosnia during the conflict is well documented with both local and international actors engaging with the black-market trade of reselling aid, such as 'food and other basic necessities to starving populations under siege in cities such as Sarajevo, Bihać, Goražde, and Tuzla' (Hromadžic 2015: 163). This often transcended spatial and conflict boundaries as it involved 'trading with the ethnic enemy or the UN soldiers' (Hromadžic 2015: 163). As observed, the impact of the war is not only evident regarding spatialised ethno-nationalistic divisions, but tensions over monetary resources, which can be linked to the conflict at different levels. J's reflection also presents the issues experienced due to the post-war economic climate of the city, which relies on starting capital as the country has transitioned from socialism to capitalism directed by the neoliberal intervention (Interview 2 2015). Capital, or access to funds remains an important factor in the capabilities of actors to transform space. In particular, this sub-division of movement crosses ethnic boundaries as the 'war profiteers' are not designated as having a particular ethno-nationalistic background (J Interview 2 2015). These persons are instead only characterised by their illegal behaviour which in turn has a negative impact on other individuals' capability to transform their narrative of the city space. In the post-war context, according to J, the majority of people are not experiencing 'a good life' (Interview 2 2015).

However, the housing situation in Yugoslavia was not egalitarian as Archer (2018: 141–142) observes white collar workers were more likely to benefit from socially owned housing than blue collar workers. Reflecting social inequalities but somewhat differently than in a capitalist context 'white collar employees and managers were significantly *less* likely to own property than poorer workers', however some workplaces subsidised the individual construction of houses (Archer 2018: 142). Archer (2018: 143) notes that 'home builders spoke about the "spirit of the time"' as a motivating factor in engaging with this process, something that can be reflected on in the broader narrative of Yugoslavian identity as a temporal moment. This is shaped by lens of nostalgia and exacerbated by the chaos, upheaval, and corruption of the war that followed. Nevertheless, socialist ideology impacted on everyday lives and individual biographies. As Koleva (2008: 28) notes we see our lives not simply through grand political events but how our own everyday experience forms our narrative, therefore our understanding of our lives is held against a 'horizon of expectations related to what a good life should be like'. This is reflected in what J (Interview 2 2015) noted about the unequal experiences in the post-conflict city, demonstrating a dissatisfaction with socio-economic barriers, reinforced by corruption. As highlighted by Moore (2013: 75) corruption in city planning and property development, is an ongoing issue with the easiest way to get a permit being, avoid official routes, and 'pay for the services of one of the city's well-connected criminal intermediaries'. Fundamentally, the capability of actors to transform space is enhanced by capital, though this is something that is time and context dependent. It can be surmised, that a 'war profiteer' may be able to buy or sell a house for themselves, however, they may face challenges rescripting their identity as a 'war profiteer' (J Interview 2 2015). Significantly, participant J does not interact with persons considered to be 'war profiteers', however, their 'nice life' make them visible to J as a distinct "other" in the city (Interview 2 2015). Notably, there is no ethnonational distinction in the identity of war profiteers as they were from different ethnic groups in the conflict and also included international actors (Hromadžic 2015: 163), this demonstrate that actions during and following the conflict have established economic sub-divisions in the city.

City Area Divisions

The institutional spatial division in the city correlates with the 1992–1995 conflict division, which did not exist before the 1992–1995 Bosnian War. While this staging legitimises the narrative of the conflict, there are open or shared spaces in the city. As previously observed, this does not mean that these spaces exist without division, but that these spaces do not actively maintain or perpetuate the ethno-nationalistic divisions. Fundamentally, the narrative of Mostar as an ethno-nationalistic divided city is accurate but also reductive. While there are political and social divisions in the city, the location of spaces on either side of the city does not always define the identity of the spaces (and therefore those who use the spaces). In focus, most spaces are not staged as 'Bosniak' or 'Croat' due to their location within the city.[4] There are notably more commercial premises in the West of the city, while there is a higher number of buildings damaged from the conflict still unused and empty in Bosniak areas. In contrast, there are several construction projects in progress on the Croat side of Mostar.[5] As a result, parts of the East of the city have been staged as derelict since the end of the war. Most notable of which is Maršala Tita (one of the main streets on the Bosniak side) which, 'looks like the war ended yesterday' (J Workshop 2015). With regard to the staging of public space, the impact of the conflict is still visible on the East side of the city. Due to the presence of abandoned and unusable buildings, there are notably fewer opportunities for development of new social spaces and, as a result, fewer incentives which facilitate social movement to these areas. Furthermore, such space has the potential to evoke conflict memories as it visually maintains the spatial destruction of the conflict.

Nevertheless, the use of space, and even the public intended use of space cannot be directed fully by the place in which it operates. For example, Mepas Mall (the largest and busiest shopping and entertainment centre in the city) is in a Croat municipality, while the Mostar Rock School is located in a Bosniak municipality. However, these spaces are not imbued with narratives of ethno-nationalistic identity, though there are a multiplicity of variables that impact on the narrative of space more broadly, such as, gender, class, religion, race and ethnicity. Fundamentally, there is a challenge to understanding all variables of movement as how we understand space and the social roles of space varies moment to moment. Therefore, unpacking the narrative of movement in spaces is an important step in order to contextualise the movement, as analysis of the location of social spaces in a divided city can overlook the meaning (if there is any overt or self-aware meaning) behind movement to the space. This is not to say that all spaces transcend the ethno-nationalistic divisions in the city, indeed some correlate with and actively reaffirm the divisions. Notably, spaces within an ethnically divisive area may be interpreted as divisive when established with a narrative or symbolic association with the identities partitioned by the conflict. For example, the 'Kosača cultural centre' (referred to as Kosača) is the Croatian cultural centre which is on the Croat side of Mostar. On the same side of the city is the Old Glass Bank (also called variations of this such as, Glass Bank or Old Bank), while this is also on the Croat side of town, it is not scripted or staged as a "Croat space" as nationalistic symbols are not as part of its visible spatial identity. The Old Glass Bank, among other spaces such as the Rock School and Mepas mall, demonstrate that ethno-nationalistic areas do not determine the staging or scripting of space as either Bosniak or Croat. This is important for analysis but also discourse surrounding the city which is intrinsically relevant to the conceptualisation of Mostar as a divided city, and to the narrative of other divided cities. Fundamentally, the capability of social actors to transform space applies to all cities not simply those categorised as divided cities. And while the naming of divided cities is useful in drawing attention to exacerbated issues in such cities, it sometimes can produce a limited scope of analysis. The institutional division of the city in Mostar is of course causal to the definition of the space as a divided city, as highlighted, this does permeate the city space though the impact of this on movement cannot be considered a universal experience for residents;

> I speak Bosnian but I often go to the other side of the river, I spend more time on this side [participant indicated the Croat side] but they have a lot of great cafes.... (A Interview 2015)

The above quote demonstrates the linguistic differentiation of the two sides in the city. Participant A describes themselves as a Bosnian speaker who crosses the river, which links spatial divisions with linguistic divisions, despite the linguistic similarities of the Bosnian and Croatian languages. However, the qualifying information about movement to the other side of the river is a self-positioning statement that the ethno-nationalistic divide does not impact on their movement. Nevertheless, through the linguistic differentiation, ingrained conceptualisations of the spatiality can be observed. When asked directly about the divisions in the interviews, all participants expressed an awareness of the divide in some respect;

> *I*: I don't think about divisions
> *J*: I don't see how it's not divided. (Workshop 2015)
> *A*: sometimes I can notice [that the city is divided] but I know a lot of different people but [sic] I am not judging them, it doesn't bother me. (Interview 2015)

Through this A illustrated that they are conscious of the ethno-nationalist division in the city, though they considered their movements to be unaffected by this division. Significantly, in their everyday schooling, A attends a 'two schools under one roof' modelled college, which is on the Croat side of the city. However, A moves on both sides of the city as they live on the Bosniak side and also attend the Rock School. Through the above quote, A distanced themselves from the discourse of the eth-no-nationalistic division in the city and elaborated that 'My Aunt is a Catholic but we are all Muslims, it is not a problem' (Interview 2015). While of similar ages to A, participants' D and E had different educational experiences through attending the UWC, but have comparable stories of transformed spatial movement. In reflecting on the effect of his attendance at the school participant D noted that prior to attending the college;

> I had a lot of friends, but I didn't know it, everything was on this side [Bosniak], but now I go with my friends [to the other side] it's the same either side. (Interview D and E 2015)

Through this D's knowledge of the institutional staging of the spatiality is demonstrated in that there are 'two sides' to the city. While D's social movement previously aligned with the institutional staging of the city space, through their attendance at UWC they have experienced new movement across the city. The change in spatial perception reported by D demonstrates how perceptions of staged space can be socially rescripted through movement in spaces of peace or shared spaces (such as the UWC). Furthermore, this demonstrates the influence of other social actors in rescripting space, in this example, school friends. However, while movement has increased for D, it does not mean that the whole narrative of the city is transformed as D noted that 'sometimes we are restricted in where to go, in some places I am from the "other side" in some bars' (Interview D and E 2015). However, this does not fully deter movement as D observed, 'there are more opportunities to socialise now, every other friend is from the Croat side' (Interview 2015). Fundamentally, a core variable which impacts upon D's movement is the presence of other social actors who categorise and label him as distinctly from the '"other side"', and therefore out of place in some spaces within Croat city areas (Interview D and E 2015). Fundamentally, D observed that divisions do not impact on his movement summarising that, 'I go to most places, but I don't like some neighbourhoods, but places I am familiar with, that's okay. We were a country brainwashed'

(Interview D and E 2015). In this reflection, participant D positioned themselves as a person who does not perpetuate the ethno-nationalistic divide through movement and reflected on the instrumentalised conflict division which directed and directs the perception of space (D and E Interview 2015). Critically, this narrative shifts the responsibility of the perpetuation of the conflict division from social actors who use the city space, on to the institutional actors who have staged it (Interview with D and E 2015).

Some participants noted that the institutional division did not impact on their movement at all. In consideration of the divisions in the city, participant C observed, '[i]n my mind [the city] is not divided and I wonder why people don't cross' (C Interview 2015). This perspective is evident in C's map of which shows movement across the city. Moreover, through this statement C positioned themselves as a person whose movement is not restricted by ethnic divisions, but that they are aware that many people do not cross the city. As a middle-aged woman who teaches at UWC, it is likely that participant C would not use the same spaces as participant D, a young, male student. This demonstrates that our positionalities are produced through unique spatial trajectories which impacts on rules and coded modes of interaction we are expected to perform, or to not perform. Our spatial trajectories also impact on how we may move through space, as C qualified their limited (in that it was mostly city centre-based) movement through noting that that 'I don't move too much [in the city] because I don't have time' (C Interview 2015). This element of functionality was also echoed by participant I who observed transport limitations impact on their movement 'If I drove or had friends in different spaces I would move around a lot more' (Workshop 2015). This reflects a close-distance to spaces, wherein an individual does not have the time to use space or the resources to travel to it; as Soja (1989: 121) observes 'social life is never entirely free of such restrictive impingements as the physical friction of distance'. Fundamentally, distance does not have to be spatial, it can be more of a temporal issue (and this is of course relative), for example moving around a city could be considered too time consuming without personal transport. Critically, in I's narrative the non-movement involves a lack of resources but also a lack of friends in other spaces in the city, which does not implicate divisive movement, though it may establish divisions.

An irreverence of the city divisions was discussed by participant E who noted that; 'I live in the Bosniak part, I don't go to the Croat part, I am not afraid. I am Serb I don't care' (Interview D and E 2015). Through this, E provided a self-identification of his ethnic identity, which includes the expectation of how this impacts on movement in the divided city. By qualifying non-movement through their ethnic identity as Serb, E categorised himself as a spatial "other" in the city, and therefore unaffected the Croat-Bosniak division due to his ethnicity (Said 1979). Following this, D added;

> We [Croat and Bosniak] both hate you (Both laugh.) That is a joke, just a joke. (Interview with D and E 2015)

The joke reflects the dynamics of ethnic divisions of the 1992–1995 conflict in BiH, and also demonstrates the use of humour as a way to negotiate transgenerational narratives. The use of humour indicates a social rescripting of the divisions, as friends D and E co-opt the narrative of the conflict divisions. The repeated qualifying statement that it was 'just a joke' is indicative of an awareness of the participants, of not only the conflict divisions but of how myself, an international researcher may interpret this exchange (Interview with D and E 2015). As Zelizer observes humour serves a functional role in transforming conflict relations and functions as 'a coping tool' (2010: 3), or to relieve tensions. Additionally, Plester (2009: 98) notes it can also be used to soften 'difficult messages [or in] addressing painful instances'.

Critically, while ethno-nationalistic divisions do not dictate all movement, it is notable that there is an awareness, and for some a potential threat, of maintained divisions. However, notably involvement in shared spaces has increased some participants' social movement, which has rescripted their perceptions of the divided city space and some social actors therein. While movement to shared spaces does not equal a panacea for a city divided by ethno-nationalistic narratives, evidently the act does hold potential to transform social relations, not only in those spaces but in the wider city scape.

DIVIDED EDUCATION

The divided political system models the education system in BiH and as such maintains ethno-nationalistic divides. This means that multiple collective narratives of the space are maintained and taught which attempts to direct the social use of space through a manipulation of identities. In

BIH divisions are entrenched through multiple divergent narratives of the 1992–1995 war which seek to maintain the divisions of the conflict through establishing blame, responsibility, and legitimacy;

> In Bosnia there is always five stories, Bosnian, Croat, Serb and other, and then the real story. If you talk with some Croat people they will tell you that we (Bosniaks) attacked them, I know in my people there was really bad people [sic]. But everyone knows what happened here.... They [Croatians and Serbs] were ready for wartime. We were not. We were attacked, we had to defend. Ask anyone. (K interview 2015)

As demonstrated in participant K's narrative there is an awareness of the multiplicity of narratives of the conflict but also an intent to tell their view point of the story of the war, though K is reflective in noting that the experience during the war was not uniform for all ethnonationalities (K Interview 2015). In 2008, Open Society Foundations published some translated excerpts from textbooks used in Bosnia-Herzegovina; which includes territorial, historical, and social claims of superiority, including 'Islam is the best religion; Orthodox Christianity is the most important religion, Muslims are Islamic Serbs while Croats are Serbs Catholics, [and] Bosnia and Herzegovina is a centuries-old Croat state'[6] (Alic 2008). Through a divided education system which involves different history lessons, alternative historical, and as demonstrated, divisive narratives, are inculcated into youth through education. The formation of curricula is set up around the three main ethnic groups, which is divisive through these narratives. However, it is doubly divisive as there are limited provisions for minority students such as Roma, this replicates the exclusion of Roma politically in BiH (Alic 2008). In Mostar there is a drop-in centre for children 'who live or work on the street', these children can use the centre for food, hygiene services, and some educational support (N, P Group Interview 2015). The children who use the centre are generally Romawho are largely excluded from the BiH education system, and as a result, there is identifiably a 'low level of education' among ethnic Roma in the country (Bosnia-Herzegovina, Presidency Plan 2014: 2). Due to the vulnerable economic position of Roma, children are often forced to work meaning that they often miss out on education opportunities, 'they want to go to school but their parents resist it' (Q Group Interview 2015). The situation is significantly worse for some Romagirls who face multiple barriers in accessing education and a stricter patriar-chal control on their movement and use of space in the city;

> It is worse for girls. Mainly boys go to school, if any do – it's usually boys. They are married very early especially girls, mainly minors, but they live together and have kids. [...] The women and children beg only, not hus-bands and older men, they are "not used to it". I was reading a book about a Princess with a little girl and I asked the little girl were does the princess go? And she replied, "The princess goes to beg." This is how they live. (O Group Interview 2015)

In order to access the daily centre services, the children require permis-sion from a parent, if they are able to attend they frequently have diffi-culty concentrating and engaging with learning activities (N, O Group Interview 2015). Roma children who have the opportunity to attend mainstream school typically come from settled Roma families, whereas those who come from a traveller Roma family are less likely to be in education (O Group Interview 2015). Critically, Roma are maligned globally and face stereotyping and scapegoating (Fox and Brown 2000: 145). In particular, traveller Roma, due to their spatial fluidity, are uniquely othered (Said 1979). Fundamentally, place affiliation and links to place influences where and how frequently individuals can move, it ties persons to place (creating 'place identity') and can serve as a for-mal source of identification through citizenship and institutional docu-ments (Proshansky et al. 1983: 58). Traveller Roma can be considered 'the [*placeless*] other', wherein they are securitised due to a lack of place identity (Said 1979). This can be theorised as stemming from the spatial reliance social actors have on place as a construct of identity. However, in BiH the marginalisation of Roma is, in part, a product of the three-pronged power structure instilled through the Dayton Agreement which entrenched the division of the 1992–1995 war and excludes ethnic minorities.

As a space of exclusion for minorities, the educational system is observably also a space of incubation for divisive political narratives; as Björkdahl (2015: 121) observes the 'two schools under one roof' model of education facilitates a 'spatial, cultural, and linguistic separation of the two communities by the international community for conflict manage-ment purposes' (Björkdahl 2015: 121). Notably, such schools are staged for a specific set of actors, with schooling typically taking place in shifts in order to maintain divisions. Due to the 'temporal dimension' of the spatial division (Björkdahl 2015: 122), the shifts mean that the students are not afforded even brief opportunities to socially rescript institutional narratives. Reflectively, the enforcement of divisions, not only spatially but temporally, indicates in its absence, the potential for social interactions to disrupt the institutional narratives of the politicised education system. The combination of spatial and temporal divisions attempts to disable unintentional opportunities for rescripting taught historical narratives. Unintentional rescripting can be understood as moments of interaction, which are not motivated by a conscious decision to move and engage with the other. Such informal interactions normalise the presence of 'the other', however, the divided education system dissuades this possibility (Said 1979). Fundamentally, institutional staging disrupts rescripting opportunities twofold and leaves the responsibility for inter-ethnic interaction with individuals who must find an alternate location (outside of formal education settings) to engage in the rescripting of divisional narratives. This has a further temporal challenge due to the shift model of schooling, meaning that one group will be engaged in school while the other is 'free' and therefore leaves very little time for inter-ethnic interaction. In Mostar 'education is [by] nationality', and is pescriptive, according to J, 'with the Croat curriculum there is a choice not to go to some classes, but you couldn't not learn Croatian' (J workshop 2015). The education system draws significant criticism; in particular with regard to the involvement of religion in schooling;

> I hate our education system, they really have a problem if you're open minded if you don't attend church you will have trouble in school. My sister had trouble in school [as] she did not want to attend some classes. (G Interview 2015)

Critically, as previously noted, the education format of 'two schools under one roof' has faced judicial opposition, and following a municipal court ruling in Mostar, the system was abolished in 2012 due to the discriminatory nature of 'ethnic segregation in the schools of Stolac and Čapljina' (Björkdahl 2015: 121). The case was put forward by 'human rights NGO Vaša Prava (Your Rights)' (Dzidic 2014) this appeal judgement was delivered on the 27th of April 2012 and called for an 'integrated and joint curriculum by the 1st of September 2012' (Monitoring

Committee of the Council of Europe 2012: 3). While the Monitoring Committee of the Council of Europe noted that the ruling would 'probably be appealed', it summarised that 'the NGO Vaša Prava [intended] to file other suits' (Council of Europe Report 2012: 3). The municipal judgement found that two schools under one roof 'can only enhance prejudice and intolerance towards others, and lead to further ethnic isolation' (Dzidic 2014). While Mostar cantonal court claimed the expiry of the statute of limitations on the case, in November 2014, the original judgement was upheld by the Federation Supreme court which observed the 'systemic discrimination which carries on continuously' in the divided education system (Dzidic 2014). Despite such efforts the education system remains divided.

Fundamentally, the educational system provides a script for future generations. This script is in the form of different histories taught to respective ethnic groups. Furthermore, the education system stages spaces of education as divided which means that there is a lack of opportunity for school children to meet and interact across divides. While educational divisions do not represent the full extent of the divisions in the city, this does have an impact on other spaces in the city. Fundamentally, spaces can be divided ethno-nationalistically, but may also be sub-divided, as this exchange reflects;

J: No institutions do stuff for youth, this is something the city should do. City institutions are privatised to serve nationalist means.
I: A lot of facilitates are private, sports societies are divided.
J: Youth activism is potentially politically emoted.
I: There are private ballet and judo facilities and an expensive kindergarten centre. (Workshop 2015)

In the above excerpt, participant J and participant, I discuss the ethno-nationalist divisions in sports and youth activities. Notably, the variable of class impacts on the use of the spaces, through the privatisation of facilities and the resulting cost of participation within these extra-curricular settings. While this is an issue for J and I, the existence of divided facilities can be analysed as indicating a demand. As Palmberger (2013: 545) notes, in Mostar, there is a broad range of experiences and perceptions of the divisions. For some, the division is 'welcomed' for others it is 'an obstacle in the way of normal life' (Palmberger 2013: 545). Critically, the ethno-nationalistic division is profited from in such spaces, and further establishes divisions of movement. The examples given refer to facilities that would be used by youth directed by their parent or guardian. From an early age in divided cities, the education of youth depends on a triangulation of variables, the spaces they are educated in, the spaces in which they engage with hobbies, and the space of the home. Intrinsically, the space of the home is typically one of the main driving forces for the perpetuation of transgenerational narratives of division. Much like the interconnectivity of Galtung's (1990: 294) violence triangle, if the space of the home and the narratives of authority figures in that space is one that is divisive, there is unlikely to be a counter narrative taught in formal spaces of education, this is also more likely to be reaffirmed by hobby spaces as parents direct these also. With education as such a socially, politically, and legally contested space, the schools that are able to integrate students can become important models for the integration of not only the curriculum but youth.

THE UNITED WORLD COLLEGE AND GYMNASIUM

In 2004, the same year as the completion of the reconstruction of the Old Bridge (Stari Most) , and the opening of OKC Abrašević; the Gymnasium Mostar opened to all in the city. Previously regarded as one of the potential 'building-blocks, and possibly a linchpin, of a bulwark designed to seal off "nationalised territory"' for Croats, the Gymnasium now allows pupils the opportunity to interact across ethnic boundaries (Wimmen 2004: 5; Björkdahl 2015: 123). Though students are divided educationally in different classrooms, they are in the same building. In effect, this bridges a potential gap created by schooling, which as previously noted can take place in shifts, typically with a buffer time zone that creates not only a temporal but a spatial division between children of different ethnic groups. In Mostar, the opening of the Gymnasium meant that 'for the first time since 1991, students could learn together' though notably teaching spaces were initially distinctly marked as ethnically separate (Björkdahl 2015: 124). The Gymnasium is in the same building as the UWC, and though the school operates under the two schools under one roof policy, levels of cross, ethno-nationalistic interaction have improved in recent years, as observed by Björkdahl (2015). Though still functioning under the two schools under one roof principle with separate classes, internal borders have been deconstructed (Björkdahl 2015: 125). In particular, from interviews conducted in 2014, Björkdahl (2015: 125) observed that 'the spatialisation of ethnicity' was notably 'less clearly identified by symbolic and material markers, i.e. the classrooms are no longer marked' for split ethnic usage. Furthermore, Björkdahl (2015: 125) noted that inter-ethnic meetings used to occur in private spaces, such as sharing a cigarette 'during break in the bathrooms' in what is considered 'the spatial fringes of ethnic space'. However, in more recent years such meetings occur in public outside the school or in 'extra-curricular activities' (Björkdahl 2015: 126).

In 2006, the Gymnasium became the host space for the UWC Mostar, now on the third floor of the building. The UWC opened in 2006, with the directed 'aim to contribute to the reconstruction of a post-conflict society' (UWC 2015a). The school facilitates a shared space through a mixed student body sometimes ranging in age from 15 to over 19 though generally between 16 and 19 (UWC 2016a). Notably, UWC Mostar is largely international, with an estimated '30 students from Bosnia-Herzegovina and 70 from all around the world' (UWC 2015b: 1). As the UWC and the Gymnasium are in the same building, this means that students attending either institution inevitably meet and mix. Additionally, students from the UWC frequently engage in activities which seek to benefit the community, such as picking up litter by the side of the road or contributing to the SAF in the city (C Interview 2015; G Interview 2015). UWC Mostar is the first, and so far, the only UWC established to assist in 'the reconstruction of a post-conflict society' (UWC 2015a). The school operates within, but also beyond the space of Mostar. Figuratively, it is an international spatial mediator in the city and students in residences share rooms 'wherever possible [...] with students from different ethnicities and nationalities' (UWC 2016b). C is a key figure in upper management and teaching at the UWC (Interview 2015). During the interview, C noted one of the students' residences, where some of the international and national students from across BiH stay and explained that 'it was reconstructed from a ruin, so it is really connected to me' (Interview 2015). The student residence C is referring to is located on the Bosniak side of the river and would have been on the front line during the 1992–1995 conflict. The spatial transformation of the building can be considered to be reflective of the aims of UWC Mostar in post-conflict reconstruction, socially but also physically. The reconstruction indicates a capability of the school to engage in large-scale transformation in the city, this can be seen physically but also and intangibly through mixed opportunities to socialise. The reconstruction and renovation work contributes to the local space through the transformation of derelict buildings into residential property, as a result the narrative of the space is transformed. Additionally, the reconstruction of the building which C discussed and the use by students contributes to business for nearby shops and cafes across the city (Ethnographic Research, September 2015). The capability for the UWC to engage in reconstruction in the city is further demonstrated by the 2018 signing of a co-ownership agreement for the third fully owned UWC student residence in the city, the construction will transform a derelict building near the Old Bridge (UWC 2018b). Currently, the school spreads out into the city in the form of the three residential spaces, two of which the school currently owns and one it rents (UWC 2018a: 3, 2018b). Conceptually, the school also operates across the city by promoting student collaboration and involvement in different events and organisations in the city, such as the Rock School and the SAF. Through this the UWC can be observed to operate locally as part of a network of spaces. The UWC Mostar is a notable example of integrated education, which is open to international and national students, it is important to note that this is an internationally led organisation outside of the BiH education system. However, the school demonstrates the potential integration of students through education. For some participants the education system had previously limited their spheres of activity, as noted previously, attendance at the UWC increased participant D and E's movement in the city;

> D: My family are open minded; [but] I have more friends now. Sometimes we are restricted in where to go, some places I am from the "other side" in some bars. But there are more opportunities to socialise now, every other friend is from the Croat side.
> E: [Nodding in agreement]. (Interview 2015)

While for participant D, 'every other friend is from the Croat side', divisions in the city remain apparent as D is othered in some spaces (D and E Interview 2015). This dual negotiation of the ethno-nationalistic divisions demonstrates that, though ethno-nationalistic divides do not dominate all social movement in Mostar, there are social actors who maintain this divide. Some spaces are socially scripted as divisive, while attendance at the UWC increased the ethno-nationalistic diversity of the participants' social circles, and therefore movement, it is notable that this does not mean all spaces are accessible. This demonstrates the interactivity of social actors and the meaning or use of space and importantly highlights opportunities for transformation. While some youths through the UWC and Gymnasium have had the opportunity to engage across the divide it is notable that for many in the city opportunities for interaction may be limited as students face politicised spaces of education in primary and secondary schooling, and in higher education.

HIGHER EDUCATION

The ethno-nationalistic divide in higher education in Mostar is spatialised as the two universities are on either side of the city and are representative of the divide in the city, with University Džemal Bijedić of Mostar (located on the Bosniak side) and The University of Mostar (located on the Croat side). Though both institutions stem from the same establishment of a higher education institution in 1978, the universities now exist as separate entities in the city. Two of the research participants had previously performed the role of Student Union presidents at the each of the universities in the city and held office at the same time. When in office participants M and K organised shared activities;

> We tried to do some stuff together with [the] other student's union, like concerts and friendly football matches. They were first projects of that kind in Mostar, unfortunately, no one before us did that. People are not really interested in social projects, that's the biggest problem now I think. You can't learn [sic] [people] anything if they don't want to learn. (K Interview 2 2015)

These activities were typically organised to occur in the city Centre to encourage involvement from both universities between the sports teams and always had a '[g]ood spirit and competition' (M and K interview 2015). In this respect, through the student unions, the divided higher education system became a network for fostering spaces across the city which potentially transformed the divisions of the conflict through shared movement and use of space. Following on from their positions in the respective student's unions; participant K has become a youth politician, and participant M has helped established a local NGO. As small-scale initiatives had previously been important for both M and K's movement and use of the city, I asked how they saw the potential of youth centres to change the city through shared activities;

K: Youth centres are cool, but we need economy, we are talking about serious stuff.

M: We are not young we are talking economics. (Interview 2015)

Critically, the core contemporary issue in the city identified by the participants, was that of economics; when the two participants headed the student unions, priority projects involved the establishment of youth based interactive activities, for the betterment of the city. Though perhaps, at the time, personally benefiting from these circumstances, through a change of career, a shift of focus is identifiable. This demonstrates the influence of space and positionality, as that which influences not only an individual's sphere of influence but also personal perspectives and objectives. The stage an actor engages with, can support (but may also direct) their capabilities to rescript space. In focus, the example of M and K's student union work in the city presents the space-time dependency of actors' interests, but also how this can facilitate rescripting of space and social relations within space. This demonstrates the capability of individually driven cooperation in the ethno-nationalistically divided higher education system. However, some challenges of the current higher education system are hard to socially negotiate, and issues regarding a lack of resources and corruption were noted by several participants;

The cost of school is an issue but also how they construct their syllabus around their [the professors] books. My brother and his wife are students, she is studying economics, ICTs and economics, they need programs with economic parameters and they don't have a computer so she works in a notebook. (J Interview 2 2015)

A similar issue was raised by participant K who came to the interview with specially printed paper he was taking to his partner to use for calculations which should have been done on a computer. Though a lack of resources, and the tendency for courses to be constructed around a course leader's book, are not exclusive to BiH, these issues nevertheless impact on the environment created for learning. There are identifiably different types of corruption in the education system which also operates along ethno-nationalistic narratives;

Nothing is independent in this country, so education is not exempt from that. First off, all the persons appointed to decide on the educational matters are politicians concerned only by pushing their agendas. This is most seen [sic]in the examples of curricula and textbooks. Then politicians will also install certain people to their positions in the education, such as rectors of the University, professors, presidents of students' unions et cetera. (J Interview 2 2015)

Critically, politicians are intrinsically involved in the construction of curricula and in maintaining power in higher education facilities through cronyism. The widespread presence of corruption in BiH is also discussed by the non-governmental organisation Transparency International which aims to fight corruption. In a study conducted, from 'October 2011 to February 2012' involving '2,000 students and 500 university employees' in BiH, both groups reported the presence, and issue of corruption in higher education institutions (Transparency International 2013: 191). In total, 56% of students surveyed identified 'corruption as a dominant feature of Bosnia-Herzegovina's education system', which was listed third as a core issue behind 'the lack of workspace and the neglect of practical knowledge and skills in the curriculum' (Transparency International 2013: 191). Additionally, 61.4% of staff surveyed, though aware of corruption, believed that 'it occurs as isolated cases in which a relatively small number of people participate' (Transparency International 2013: 191). What is apparent is that corruption crosses the ethno-nationalistic division. Therein, the capability for individuals to unite against political corruption reflects the potential for conflict narratives to be traversed. The unifying potential of protesting political elites can be observed in the 2014 protests, as previously noted, with International Crisis Group regarding the protests in Mostar 'as a joint Croat and Bosniak affair' (ICG 2014: 4). The influence of political staging in higher education is nevertheless evident, as through maintained top-down influence, these narratives have become 'embedded in the curriculum and what the students learn is the politicians' version of events, history, language and literature' (J Interview 2 2015).

However, the influence of the higher education system does not operate in isolation and is facilitated by earlier schooling. In particular, if young adults have grown up attending school in the divided education system they may have had little chance to meet and interact across the conflict divide. Politically, the divisions are instrumental in maintaining an electoral pool, this is demonstrated through a primordialist account by Ivo Miro Jović, a Bosnian Croat member of the state parliament;

the divisions don't start at school, but at home. We are being raised differently from the day we are born. A man is what he is - he is being raised by his family who instil in him a sense of belonging to his community. That should not insult anyone, just as different curricula for Bosniak and Croat children should not be a problem to anyone. (Kamber 2011)

Ivo Miro Jović's statement exemplifies not only an attempted normalisation of educational divisions, but the perception that ethno-nationalistic divisions are part of every citizen's socialisation. Indeed, as previously observed the space of the home is crucial for peace or conflict transformation, and divisional narratives may be part of a child's upbringing, however, these are politically engineered. What is notable is that political divisions are indoctrinated at an early age through the staging of the education system by institutional actors to direct movement and maintain spatial divisions. This has a transgenerational impact which crosses space and time, from the public to the private. Fundamentally, this demonstrates the capability of social actors to traverse divides, as the ethno-narrative of divisions is established from an early age in order to inculcate social actors into divisional thinking. The ethno-nationalistically staged space of education aims to direct movement by limiting exposure to the spatial other, which in its absence, demonstrates its potential power.

Division Spatialised

What can be said about the visualisation and spatialisation of the conflict division in the city of Mostar is that there are multiple shifting definitions of space which are fluid, interconnected and socially responsive. In Mostar, a conflict of ideologies is observable. The city is politically and jurisdictionally divided, but there are also social divisions which can be seen to transcend ethno-nationalistic identity, while some script the city as divisive, others attempt to rescript the institutionally staged space as shared. Street art and graffiti in the city are spatial indicators of peace and conflict respectively, with the SAF conceptualised as a temporal and physical space of peace through the festival each year, and the art produced, which remains following the festival. Notably, the location or density of street art in the city correlates with the location of open spaces in the city, namely OKC Abrašević, the UWC, and The Old Glass Bank. At the site of the Old Glass Bank, the "message" of the SAF Mostar is visible through the artwork. As the SAF is a socially led initiative, the art can be observed to be a social narrative of not only the space but the expectation of the space. Furthermore, the example of the Old Glass Bank demonstrates that social use of space has the potential to alter the staging of space. Through the artwork located at the bank, the derelict building is rescripted into an open-air gallery of street art. As the festival was established by several OKC Abrašević members this demonstrates the capability of social actors to facilitate shared spaces in the divided city.

However, visual signifiers can also demonstrate the divisive nature of space, and the example of street art can be juxtaposed with Ultras graffiti, stickers, and the fascist graffiti in the city. Both street art and informal graffiti provide markers that set out the narratives of the spatiality and aim to direct the movement and use of the space, which in the case of fascist graffiti and stickers may be regarded as divisional and threatening in nature, though it must be noted that anti-fascist graffiti counters fascist graffiti in the city. With regard to divisional graffiti, the visibility of this demonstrates that social rescripting can challenge or correlate with institutional staging.

Fundamentally, the maintenance of administrative divisions is visible through city areas. However, participants noted that the institutional divisions do not act as a boundary for movement or influence their use of the city space, nevertheless, participants did sometimes reference movement in the city as either being "one side" or "the other side". Notably, the participants joked about ethno-nationalistic relations, and reported an increase in inter-ethnic friends (and cross divide movement) which demonstrates the important role shared spaces perform in the city. However, institutional divisions and corruption in education is an ongoing issue, and divisions in the education system are used by politicians to maintain ethno-nationalistic divisions (J Interview 2 2015; Transparency International 2013: 191). For many children the divided education system establishes borders and boundaries in their everyday lives, and the ethno-nationalistic divisions in the education system remain a key variable in the transgenerational maintenance of conflict divisions. This is particularly the case for Roma, for whom the education system is often spatially, and politically inaccessible. The divided education system notably aims to stage space as divided through historical narratives, however, is it observable, through participant narratives that exposure to the divided education system can be transformed by movement and use of shared spaces. Therefore, through establishing shared space or through movement and use of the spaces, social actors can be considered to have rescripted spaces in the city. Reflecting on the impact of movement in spaces of peace or shared spaces, the capability for social actors to rescript space, is enhanced through family and social circles. Holistically, institutional divisions do not always translate to a social stasis, and participants, with a little help from new friends, have been able to negotiate and rescript the staging of ethno-nationalistic divisions in the city.

NOTES

1. This is in acknowledgement that even the narratives included cannot be considered comprehensive.
2. Ethnographic research June 2014, March, April, and September 2015.
3. Ustaše refers to individuals who are members of the Croatian Revolutionary Movement, founded by fascist Ante Pavelić in 1929, who committed 'a massive genocide of Serbs, Jews and Roma' during the Second World War (Mojzes 2011: 54, 18).
4. This also applies even if they are ethno-nationalistically staged, as social actors can facilitate a transformation of the use of space, though this can be considered to be momentary and socially dependent.
5. Alongside the reconstruction of the Old Bridge, the 2005 UNESCO nomination proposed the reconstruction of several properties in the city, including Hotel Ruža (UNESCO 2005: 31), which now in the UNESCO zone and is projected to open in 2019 as a five-star hotel.
6. From Bosnian, Serb, and Croat, geography text books respectively.

BIBLIOGRAPHY

Alic, A. (2008). *Bosnia and Herzegovina: Teaching Intolerance*. Open Society Foundations. Available from: https://www.opensocietyfoundations.org/voices/bosnia-and-herzegovina-teaching-intolerance. Accessed Dec 2017.

Archer, R. (2018). The Moral Economy of Home Construction in Late Socialist Yugoslavia. *History and Anthropology, 29*(2). https://doi.org/10.1080/02757206.2017.1340279.

Bell, C. (2009). Sarajevska Zima: A Festival Amid War Debris in Sarajevo, Bosnia-Herzegovina. *Space and Culture, 12*(1). Available from: http://sac.sagepub.com/content/12/1/136.full.pdf. Accessed 29 Nov 2016.

Björkdahl, A. (2015). 'Two Schools Under One Roof': Unification in the Divided City of Mostar. In A. Björkdahl & L. Strömbom (Eds.), *Divided Cities, Governing Diversity*. Lund: Nordic Academic Press.

Bosnia and Herzegovina Presidency Plan June 2014–June 2015. *Decade of Roma Inclusion 2005–2015*. http://www.romadecade.org/cms/upload/file/9294_file3_decade-bih-presidency-plan-for-isc–final.pdf. Accessed 12 July 2014.

Council of Europe. (2012, September 5). *Honouring of Obligations and Commitments by Bosnia and Herzegovina.* AS/Mon (2012) 18 rev. Available from: http://assembly.coe.int/CommitteeDocs/2012/amondo-c18rev_2012.pdf. Accessed 27 June 2016.

Dauphinee, E. (2007). *The Ethics of Researching War: Looking for Bosnia.* Manchester: Manchester University Press.

De Certeau, M. (1984). *The Practice of Everyday Life.* Berkeley: University of California Press.

Dzidic, D. (2014, November 4). *Balkan Transitional Justice. Bosnia Federation Rules Against Ethnically Divided Schools.* Balkan Insight. Available from: http://www.balkaninsight.com/en/article/bosnian-federation-court-rules-against-school-discrimination. Accessed 4 Nov 2014.

Fox, J., & Brown, B. (2000). The Roma in the Post-communist Era. In T. R. Gurr (Ed.), *People Versus States: Minorities at Risk in the New Century.* Washington, DC: US Institute of Peace Press.

Galtung, J. (1967, September). *Theories of Peace: A Synthetic Approach to Peace Thinking.* Oslo: International Peace Research Institute. Available from: https://www.transcend.org/files/Galtung_Book_unpub_Theories_of_Peace_-_A_Synthetic_Approach_to_Peace_Thinking_1967.pdf. Accessed 25 Nov 2014.

Galtung, J. (1990, August). Cultural Violence. *Journal of Peace Research, 27*(3). Available from: http://www.jstor.org/stable/423472. Accessed June 2017, 2016, P. 12.

Hopkins, V. (2015). The Town Dayton Turned into a Garbage Dump. *Foreign Policy.* Available from: http://foreignpolicy.com/2015/11/21/the-town-dayton-turned-into-a-garbage-dump-bosnia-mostar-anniversary/. Accessed 20 June 2016.

Hromadžic, A. (2015). *Citizens of an Empty Nation: Youth and State-Making in Post-war Bosnia-Herzegovina.* Philadelphia: University of Pennsylvania Press.

International Crisis Group. (2014, July 10). *Bosnia's Future. Europe Report No. 232.* Available from: https://d2071andvip0wj.cloudfront.net/bosnia-s-future.pdf. Accessed 20 Sept 2016.

Kajan, S. (2016, September 18). Citizens of Mostar: Fascism Resides Here. *Al Jazeera.* Available from: http://balkans.aljazeera.net/vijesti/gradani-mostara-ovdje-stanuje-fasizam&prev=search. Accessed 20 Jan 2018.

Kamber, A. (2011, May 3). *Segregated Bosnian Schools Reinforce Ethnic Division.* Balkans TRI Issue 690. Available from: https://iwpr.net/global-voices/segregated-bosnian-schools-reinforce-ethnic-division. Accessed 20 May 2016.

Kaneva, N. (2011). *Branding Post-communist Nations: Marketising National Identities in the "New" Europe.* New York and Abingdon, Oxon: Routledge.

Koleva, D. (2008). 'My Life Has Mostly Been Spent Working': Notions and Patterns of Work in Socialist Bulgaria. *Anthropological Notebooks, 14*(1).

Available   from: http://www.drustvo-antropologov.si/AN/PDF/2008_1/
Anthropological_Notebooks_XIV_1_Koleva.pdf. Accessed Feb 2018.

Massey, D. (1991, June). A Global Sense of Place. *Marxism Today*. Available
from: http://banmarchive.org.uk/collections/mt/pdf/91_06_24.pdf.
Accessed 20 June 2016.

Mišna, D. (2016). *Shake, Rattle and Roll: Yugoslav Rock Music and the Poetics of
Social Critique*. New York and Abingdon, Oxon: Routledge.

Mojzes, P. (2011). *Balkan Genocides: Holocaust and Ethnic Cleansing in the
Twentieth Century*. Lanham, MD and Plymouth, UK: Rowman & Littlefield.

Moore, A. (2013). *Peacebuilding in Practice, Local Experience in Two Bosnian
Towns*. Ithaca: Cornell University Press.

Naef, P., & Ploner, J. (2016). Tourism, Conflict and Contested Heritage in
Former Yugoslavia. *Journal of Tourism and Cultural Change, 14*(3). https://
doi.org/10.1080/14766825.2016.1180802.

Newman, D. M. (2014). *Sociology: Exploring the Architecture of Everyday Life*.
Los Angeles: Sage.

Palmberger, M. (2013). Practices of Border Crossing in Post-war Bosnia and
Herzegovina: The Case of Mostar. *Identities: Global Studies in Culture and
Power*. Available from: http://www.tandfonline.com/loi/gide20. Accessed
20 Oct 2016.

Participant A. (2015). Interview in Mostar.

Participant C. (2015). Interview in Mostar.

Participant D and E. (2015). Interview in Mostar.

Participant G. (2015). Interview in Mostar.

Participant G. (2016). Interview 2, follow up over email.

Participant H. (2015). Interview in Mostar.

Participant I. (2015). Workshop in Mostar.

Participant J. (2015). Interview 1 in Mostar.

Participant J. (2015). Workshop in Mostar.

Participant J. (2015). Interview 2, follow up over email.

Participant K. (2015). Interview in Mostar.

Participant K. (2015). Interview 2, follow up over email.

Participant L. (2015). Interview in Mostar.

Participant M. (2015). Interview in Mostar.

Participant N, O, P, Q. (2015). Group Interview.

Participant R. (2015). Email Correspondence.

Plester, B. (2009). Healthy Humour: Using Humour to Cope at Work.
Kōtuitui: New Zealand Journal of Social Sciences Online, 4, 1. https://doi.org
/10.1080/1177083X.2009.9522446.

Proshansky, H. M., Fabian, A. K., & Kaminoff, R. (1983). Place-Identity:
Physical World Socialisation of the Self. *Journal of Environmental Psychology,
3*(1), 57–83.

Said, E. W. (1979). *Orientalism*. New York: Random House.

Soja, E. W. (1989). *Postmodern Geographies. The Reassertion of Space in Critical Social Theory.* London and Brooklyn: Verso.

Sontag, S. (1979). *On Photography.* Harmondsworth: Penguin.

Tonkiss, F. (2005). *Space, the City and Social Theory: Social Relations and Urban Forms.* Cambridge: Polity.

Transparency International. (2013). *Global Corruption Report: Education.* New York and Abingdon, Oxon: Routledge.

UNESCO. (2005, July 15). *The Old Bridge Area of the City of Mostar.* World Heritage Scanned Nomination. Available from: http://whc.unesco.org/uploads/nominations/946rev.pdf. Accessed 14 June 2014.

United World College. (2015a). *Who We Are.* UWC Mostar. Available from: http://uwcmostar.ba/who-we-are/. Accessed Oct 2015.

United World College. (2015b). *United World College Mostar, General Information.* Available from: http://uwcmostar.ba/wp-content/uploads/2015/12/general-information-2015.pdf. Accessed 20 Oct 2015.

United World College. (2016a). *United World College Mostar, Students.* Available from: http://uwcmostar.ba/join-us/students/faq/. Accessed Apr 2018.

United World College. (2016b). *Residential Life.* Available from: http://uwcmostar.ba/living/residential-life/. Accessed Apr 2018.

United World College. (2018a). *United World College Mostar Newsletter No. 48. School Year 2017/2018.* Available from: http://uwcmostar.ba/wp-content/uploads/2018/01/newsletter_48_WEB.pdf. Accessed Apr 2018.

United World College. (2018b, April 20). *United World College Facebook.* Available from: https://www.facebook.com/uwcmostar/. Accessed 20 Apr 2018.

Volčič, Z. (2007). Yugo-Nostalgia: Cultural Memory and Media in the Former Yugoslavia. *Critical Studies in Media Communication, 24,* 1. https://doi.org/10.1080/07393180701214496.

Wabl, S. (2017, 10 October). *Return to Europe. Kiss the Bottle.* Available from: http://kissthebottle.org/velez-zrinjski-mostar/. Accessed 20 Feb 2018.

Waende Süedost. (2011). *Gigo Propaganda.* Artist. Available from: http://www.waende-suedost.de/index.php?id=38&L=1. Accessed 30 May 2016.

Wimmen, H. (2004). Territory, Nation, and the Power of Language: Implications of Education Reform in the Herzegovinian Town of Mostar. *GSC Quarterly, 11,* 1–21.

World Bank. (2017, March). *Unemployment, Youth Total (% of Total Labour Force Ages 15–24) (Modelled ILO Estimate).* International Labour Organisation. ILOSTAT database. Available from: https://data.worldbank.org/indicator/SL.UEM.1524.ZS. Accessed Mar 2017.

Zelizer, C. (2010). Laughing Our Way to Peace or War: Humour and Peacebuilding. *Journal of Conflictology, 1*(2). Available from: http://dx.doi.org/10.7238/joc.v1i2.1010. Accessed Feb 2018.

CHAPTER 6

Social Constellations of Transformation: *Space, Place and Transformation*

The dexterity of space is not simply in its use or design but in the flexibility of the discussion surrounding what space is. The chapter focuses on the narratives of social actors and the popular spaces they use and move through. Through movement social actors transform the scripting of physical spaces. This, of course, happens every day in all cities and spaces. All space can be understood as an ever-expanding constellation of transformation. The form this takes is largely dependent on the actors using and scripting the narrative of the space and can be through signs and symbols or through movement, relations, and interactions with, and in, space. The maps below are examples produced by participants. Figure 6.1 is drawn South to North (left to right) while Fig. 6.2 is drawn West to East (left to right); this corresponds to the Croat side of the water (the West) and the Bosniak side (the East). These two maps were two of the most detailed maps produced and contain most of the spaces mentioned by participants. What is identifiable from all of the maps produced is that there are more spaces of popular of social movement in the West compared to the East of the city. While this is important to note, the maps should not be analysed in isolation as an indication of intentional divisive movement in the city. As previously discussed, there is a tendency for international researchers to search for 'political agency' in what is potentially 'routine' movements (Kappler 2013: 130). This is where the narrative of movement helps unpack the purpose, feelings, and sensations of space (if there are any), in order to negate the application of agency to functional movement. In what follows the narrative of the use

© The Author(s) 2019   155
S. Forde, *Movement as Conflict Transformation*, Rethinking Peace
and Conflict Studies, https://doi.org/10.1007/978-3-319-92660-5_6

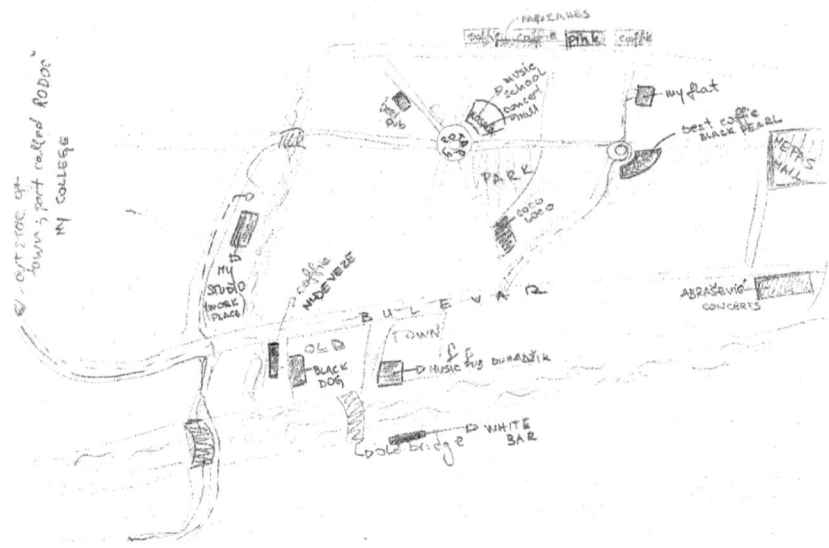

Fig. 6.1   An example of a participants' cognitive map

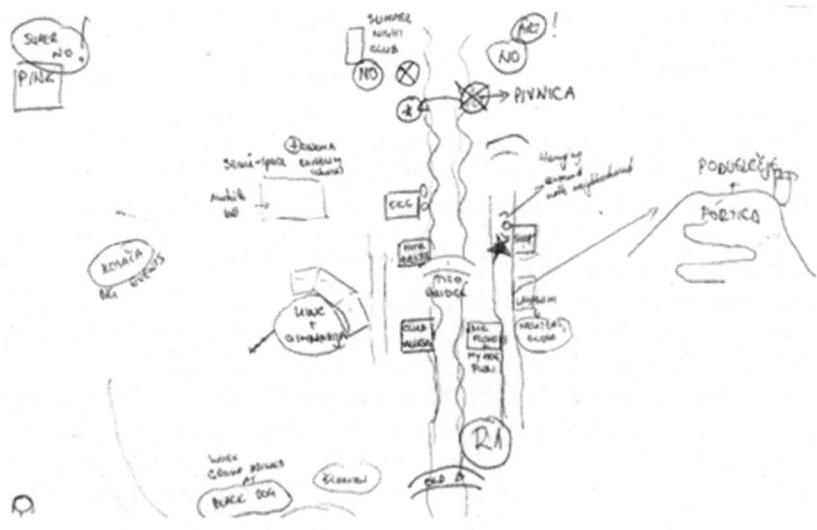

Fig. 6.2   An example of a participants' cognitive map

of different social spaces in Mostar, travelling North in the city, is discussed comparatively with the staging of the space.

THE ROCK SCHOOL

Mostar Rock School was established in 2012 and is an affiliate of the Muzički Centar Pavarotti (Pavarotti Music Centre (PMC), established in 1997) both spaces seek to transform the narratives of the conflict. The charity War Child UK founded the PMC, in what was an old girls school and was supported by, among other celebrities, 'Luciano Pavarotti, Brian Eno, [and] members of U2' (PMC 2016). The aim of the centre was to engage local children with music in the city, to 'bring relief to the people of BiH, especially to youngsters' (PMC 2016). Initially, PMC had two main goals, the first, was the use of music as something which transcended the divisions in the community and the second;

> to bring clinical music therapy as a new approach to traumatised children and young people of BiH, especially in [the] Mostar area (statistically the largest number of war traumatised children and young people were in Mostar) [this] was realised by setting up the first clinical music therapy department in the Pavarotti Music Centre. (PMC 2016)

The PMC centre can be regarded as a changing or adaptive space, as the goals of the centre are responsive to the needs of the residents of Mostar. Representative of this, the contemporary goals of the Pavarotti centre are tailored around cultural activities which are widely inclusive, through offering;

> different educational and creative activities and training to children and young people adapted to their age and abilities applying new reformed educational methods with elements of creative therapy; to offer different artistic programs to people of Mostar...to improve and develop cultural life of local community; to offer different lectures and seminars; to offer the PMC facilities to other organisations, institutions and individuals; to develop and establish relationships and cooperation with other cultural, youth, education and social organisations locally and abroad. (PMC 2016)

The goals of the PMC still involve creative therapy but have notably broadened and importantly aim to cooperate with other local youth and social institutions. With the wider focus of the aims of the school, the maintained necessity of the organisation is evident. It is also clear that through appealing to not only youth but the general community, the centre can be seen to be enlarging its assistance in the transformation of the divided city. The recent establishment of the Mostar Rock School (herein Rock School) highlights the continued importance of providing creative social space which operates without ethnic divides. The origins of the Rock School stem back to 1998 through a 'local organisation Apeiron', it was not until 2011 that the idea was properly revisited by 'Orhan Maslo, in cooperation with the Dutch organisation Musicians without Borders' and 'the Pavarotti Music Centre' (Mostar Rock School History 2015). Funding for the establishment of the Rock School came from several international sources with preliminary work funded by the Robert Bosch Foundation, a German charitable institution, and operational activities receiving support from 'the Royal Norwegian Embassy and the United States' (Mostar Rock School History 2015). The school started to form in proper, in 'August 2012, when twenty young musicians from Mostar' took part in a summer school in Skopje (Mostar Rock School History 2015). The centre can be observed as socially established though it is supported by (international) institutional funding. After a period of uncertainty in 2016, the school secured three years of funding from the Swedish Embassy in Sarajevo (SIDA) and the USAID Support Programme for marginalised groups (Mostar Rock School, History 2018). The school provides tuition for 12–25 year-olds (though there is no strict limit) and a space to practice music and interact with other students (S Informal conversation 2014). The students are placed in ethnically diverse bands by school management, in the bands they practice thematic songs such as 90s rock and then perform these at monthly shows. Through this, the school facilitates social connections across ethnic boundaries for young Mostarians. These rock bands are not only ethnically diverse but are also international as UWC students attend the Rock School. Working in the bands ensures that the students learn to communicate with one and other (S Informal conversation 2015). In this capacity, it is important to note that this is not an entirely open space, in that, for individuals to be able to use this space (in the context of the band activities), they will most likely be interested in music in some capacity (though that may be a given) and of a younger age. The Rock School is supported in its work by the UWC in Mostar which has 'confirmed the quality' of the programmes available via the Rock School and incorporates the 'school programme in one of five core outreach activities for five of its students annually' (Mostar Rock School 2015). However, there is an economic barrier to engagement with the space as students pay 15 Euros per month for two weekly lessons 'plus unlimited [use] of the Rock school space for practice using school equipment too' this is roughly thirty Bosnian convertible marka (S Email correspondence 2016). As noted by management of the school, taking the economic situation in Bosnia into consideration, 'even 15 Euro is over the limit for some, while most of them couldn't afford to pay more and only few would be able to pay 25 Euro or a bit more' (S Email correspondence 2016). Without external funding support, the potentially higher rates of the music school would be unsustainable, and so external funding is constantly required. According to S, a lack of funding would mean that the '100 young people that are connected through [the] same interests, music, would go back where they were 5 years ago. Streets. [sic]' (S Email correspondence 2016). Two participants of the research, A and B, were students at the school. Since attending the Rock School, participant A had not only been able to explore their musical hobby and advance their skills but has also experienced more of the city. Furthermore, attending the Rock School has impacted on how they move around the city; 'I met a lot of people when I started coming to the Rock School, new people. I am more open minded because there is a lot of people who I wouldn't have met' (A Interview 2015). While the participant does not refer directly to the divisions of the conflict, they allude to the divisions, through how the school changed their social network. The change in social movement, facilitated by the school, has opened up A's social network and transformed their experience of the city; '[m]y movement increased, when I walk down the street I know everyone' (Interview 2015). This narrative demonstrates a growth of friendship groups and a hobby which has increased their confidence in movement; not just as a young Bosniak in a post-conflict divided city space, but as a young person in a growing city. It is important to note that though A was aware of divisions in the city, they felt that the divisions did not impact upon how they now use the city space, or how they interact with individuals who may be of a different ethno-nationality. This was an awareness echoed by B;

> I have friends from this side [Bosniak city areas] also since I go with them to play music and I met people like [S]. We are really the same; I drink coffee with whoever. (Interview 2015)

Through this we can see that B is conscious of the divide in the city, that attendance at the school has increased their social circle, and that other social actors influence their use of space. As B distinguished their nationality as Croatian in the interview, the above statement positions B as expanding their friendship group to include friends from the East of the city due to movement to the school (Interview 2015). Notably the language of the division is a natural descriptor, through the acknowledgement of the definition of "one side" and the "other side", though not utilised as a negative descriptor, is identifiable as a maintenance of the narrative of the conflict. Through usage, it shows an awareness of the conflict division lines and how they correspond with the actors who use the space. Through the experience of A and B, it is observable that the staging of the Rock School has established a space where local youth can meet in an environment which crosses ethnonationalistic narratives. It is important to note the impact of movement in such spaces of peace has a residual effect that crosses over into other areas of students' lives and it has increased movement and friendship groups for both A and B. In this capacity, movement to such spaces has the potential to rescript not only individual perceptions of space but also promotes social relations through movement and use of the wider city space, and not simply movement to and from the shared space.

THE OLD TOWN AND THE OLD BRIDGE

The next significant space of movement moving north in the city is the Old Town area (Stari Grad) and specifically the Old Bridge (Stari Most). While this is a popular recreational and tourist space, it is also an important 'stage of memory' for the 1992–1995 conflict (Forde 2016: 478). The Old Bridge exists as an important site in the city of Mostar, not least as the namesake of the city itself. From the construction of the bridge in 1566–1567 to the destruction of the bridge in 1993, and the reconstruction of the bridge in 2004, the bridge has been, and remains, an important space in the development of the city;

> Mostar is a city built on the Old Bridge. So, it is really important for us, as a tourist attraction, but also as a core part of this city. The Old Bridge is a soul of Mostar, and people who have seen the ruining of Old Town know how much this bridge means to Mostar and the people. (K Interview 2 2015)

In July 2004, the bridge was reopened, there was a large celebratory ceremony which hailed the bridge as a reconciling structure. However, the reception of the sometimes called, New Old Bridge was mixed. As a young child of 14 years old, participant K attended the ceremony and described it as 'really awesome' (K Interview 2 2015). While participant J, who was involved in the establishment of OKC Abrašević and older than K at the time, recalled that they attended a festival elsewhere in BiH, in order 'to escape the grand opening ceremony and crowds' (J Interview 2 2015). The bridge was symbolically staged as the reconciliation of the city but has been criticised by some as a false reconciliation (Calame and Pašić 2009). This was echoed by participants in that '[n] othing really changed because of [it]. Things just continued following their course' (J Interview 2015). Despite J's feelings, avoidance of the ceremony of the reopening of the bridge, and critique of the symbolism of the bridge, J observed that she feels overall positive about the reconstruction but that 'it is strange that it is somehow old and new at the same time. It could have been done better of course' (J Interview 2 2015). This demonstrates that perceptions of space are ambivalent even within the individual, and our experience of space is capricious. One constant we can rely on is the change and transformation of space alongside our momentary experience of space, this is important to recognise as it demonstrates the transient nature of negative, but also positive experiences in space.

The Old Bridge, previously an important crossing point in the city for trade and residents, is now largely moved across by tourists—international and national and the Old Town area is a space of touristic consumerism; 'at the moment it's commercialising the old space. They are not coming back to the old traditions but sell all these Turkish things hanging around' (H Interview 2015). As H observes there is a commercialisation of the bridge and Old Town area through the Ottoman cultural narrative of the city, though this is not exclusively due to the post-conflict narrative of the space, as the bridge has historically been a popular tourist destination. However, it is important to note that the commercialisation does not only surround the Ottoman period, while glass, plastic, ceramic and plaster replications of the Old Bridge are sold, some stalls also sell tanks and jets made from bullet casings, notes bearing Tito's portrait, and a few stalls even sell Nazi badges and patches.

Significantly, perceptions about the Old Bridge vary regarding the current use of the space as one of tourism, which reflects the importance of social movement in the rescripting of staged space. The use of the space is discussed by K who noted that he does not really spend a lot of time near the bridge as 'it is mostly for the tourists, and local people spend more time in their local communities, or local pubs' (K Interview 2 2015). While K did previously reflect on the importance of the reconstruction of the bridge for the city, it is pertinent to note that the staging of the space differs significantly from the everyday social scripting of the space. With respect to the importance of the site as a location for tourism in the city, K (Interview 2 2015) observed that;

> Basically, Stari Grad and Stari Most present a big income for Mostar. People from all around the world come to Mostar because of it, but I think this could be a lot better, in the future.

Furthermore, regarding the untapped potential of tourism to the bridge, participant M proposed charging tourists' admission to the space of the Old Town and the Old Bridge;

> Talking also about the Old Town, every big town in Europe, in its cultural part, people pay entrance, over all of Europe, they pay entrance. Even if tourists pay 1KM (Bosnian convertible marka) the Old Town could be used in a better way, this could help with debt and money owed. (M and K Interview 2015)

The bridge is said to symbolise the unification of the city, however, what is notable from this narrative is that the bridge is considered to represent an untapped source of income. In the economically vulnerable country and city, the importance of capital and economic resources cannot be overstated. The link between low economic well-being and a low quality of life in post-conflict spaces is a reoccurring issue, as previously noted by participant J (Interview 2 2015) concerning the starting capital of war profiteers affording them a 'nice life.' Economic issues are observably at the forefront of residents' minds. While there is no fee to go over the bridge or to enter the Old Town, there is an Old Bridge Museum, where visitors can go up into the tower which overlooks the bridge, but also underneath the bridge.[1]

FUNCTIONAL MOVEMENT

The Old Bridge was frequently used as a spatial indicator on participant maps, though some of the interviewees made a point of noting that they do not frequent the space of the bridge. When drawing their social map, participant C drew bridge as a point of reference in the city; 'I don't go to the Old Bridge that often; the cobbles are hard to walk on...I guess the path to reconciliation is a painful one!' [laughs] (Interview 2015). Through the use of humour, C co-opted the institutional narrative of the significance of the bridge as a symbolically reconciling structure. Similar to the exchange between D and E in discussed in Chapter 5, this demonstrates the use of humour as a tool of change and protest, that can challenge narratives and existing power structures. Hart (2007: 2) recognises the powerful role both 'humour and laughter' have in acts of social protest. The use of humour is an act or critique and one which can challenge power structures whether applied in social circles or in critique of the government.[2] For participant B, this sense of humour extends to the tourists using the birdge; '[a]nd here is the terrace, we can watch people on the Old Bridge and laugh at people taking photos and they [the café] have cakes—very important.' This reflection suggests a sense of awareness at the consumability of the site of the Old Bridge, as tourists predominantly gather to take photos of the Old Bridge and Old Town. While predominately used by tourists the Old Bridge is still, for some, a functional crossing point in the city;

> D: [h]ere is the Old Bridge, I cross almost every two days but it is not special, for me.
> E: Same for me. (D and E Interview 2015)

Participant D and E's reflection on their movement demonstrates that landmarks in the city, or spaces of frequent movement may not be imbued with any deep meaning or purpose but may be simply functional. For D, the Old Bridge provides a landmark around which the map is constructed. It is also a point that most people, including myself as an external researcher, would identify, as opposed to a local grocery store, due to the social draw of the space. While the participant notes that he crosses it frequently, it is not for an emotive or ritualistic reason, but a functional one. The addition of the qualifying statement reflects an awareness of the potential interpretation someone, particularly, an external researcher could infer from such movement. Therefore, as participant D highlighted the functionality in the movement, he demonstrated an awareness of not only the context of his movements but of the potential interpretation or politicisation of such movement.

While the Old Bridge does facilitate movement across the Neretva, there are other bridges in the city which facilitate movement and may be more functional for some. Despite including the bridge on maps some participants did not want to talk about the Old Bridge, when asked if they use the bridge participant G declared 'I love all bridges' and added that;

> [w]e [the city] also have one bridge [in progress] I'd like to see built because it connects the West side [of the city] to University Džemal Bijedić because now you must walk all the way around. That bridge is important (Interview 2015).

The bridge G referenced is further north, on the river and after being recently completed in 2015 connects the Bosniak university campus to the West of the town. It is important in the city, G notes, as it provides an immediate connection across the river to an area where people have not previously had the opportunity to cross.

"Bridging" the Divided

Historically, the destruction of the bridges in the city, during the 1992–1995 conflict, prevented movement between the two sides and placed the largely Bosniak population of the East bank in a worsening condition for their basic needs (Prlić et al. judgement 2013, vol. 3 IT-04-74-T, no. 1583). During the conflict, the Old Bridge provided a crossing point in the city (as did the other bridges before destruction) for civilians, but notably also for the Army of Bosnia-Herzegovina (ABiH) (Prlić et al. judgement 2013, vol. 3 IT-04-74-T, no. 1583). With regard to the Old Bridge, the usage by the ABiH meant that the bridge was staged as 'a military target' which led to the destruction of the bridge (Prlić et al. judgement 2013, vol. 3 IT-04-74-T, no. 1582). Legally, the usage by ABiH also meant that the bridge was a legitimate target. While the bridge may not be of everyday social importance due to transgenerational public usage, the restaging of the Old Bridge, can be regarded as a positive in the city through facilitating movement and tourism. The importance of not only the Old Bridge, but all bridges in the city of

Mostar is therefore demonstrated through facilitating movement across city though they technically connect Bosniak city areas. The functionality of such infastructure is, of course, not exclusive to Mostar but in the context of connecting communities which are divided politically and spatially, the bridges facilitate movement across the divided city. While the Old Bridge provides infrastructure in the city it also symbolises a wider culture of co-existence reflected across Bosnia. Bridges in Mostar, Sarajevo and beyond (cities such as Derry and Mitrovica) remain not just symbolically important, but also physically important to divided cities. What is evident socially, however, is that the use of the bridge is one of functionality, and though the reconstruction was important, the symbolism does not translate to everyday use of the bridge. The social scripting of the bridge therefore differs to the institutional staging, which primarily is engaged with by tourists. The space is a 'stage of memory' constructed of remembering and forgetting, as the weather-beaten UNESCO information boards note it was destroyed during the war but do not discuss the narrative of the deconstruction (Forde 2016: 467). What is evident in the UNESCO signage is the institutional commitment to recognising the separateness of the three languages Bosnian, Serbo, Croatian. The UNESCO and City of Mostar sign (Fig. 6.3) has the translation of the UNESCO inscription in Bosnian, Croatian, Serbian, and English. This looks on to the bridge, while to the right of this there is an informal memorial, a small stone which compels observers, 'DON'T FORGET' (Fig. 6.4). There is another small stone on the other side of the bridge which reminds users of the space again, 'DON'T FORGET '93', this was the year the bridge was destroyed by HVO shelling (Fig. 6.5).

These stones provide a visibility to the conflict in the space, as Björkdahl and Selimovic (2015: 328) observe the 'DON'T FORGET' stone Fig. 6.4 'holds symbolic value and as such it is contested'. This was placed by 'a member of the army of BiH's 'cultural brigade' as a protest against the war'. As a result of the origin of the stone and the time period in which it was placed Björkdahl and Selimovic (2015: 328) consider the stone to carry 'a specific local provenance and meaning'. While there are complex power dynamics in the linguistic dominance of English, this can be co-opted and redeployed by local actors to their own ends and demonstrates the global-local nature of space. However, these stones are not the only narrative of the space and there is a lesser-photographed stone plaque (Fig. 6.6) which sits on a piece of the original Old

Fig. 6.3 Old Bridge UNESCO sign overlooking bridge (Photo taken by author April 2015)

Bridge on the banks of the Neretva which recounts the destruction of the bridge.

The plaque (Fig. 6.6) attributes the destruction of the bridge to the HVO forces and one which was condemned by all. It reads, 'Extremists HVO and HV destroyed the 427 year old bridge. The entire world condemned this inhumane act' (Kontra Press 2013). The plaque and rocks, alongside the UNESCO signage demonstrates that the space of the Old Bridge, is one of both remembering and forgetting. But without a significant appeal of the space outside of the tension between remembering and forgetting, the space is not one that encourages local shared usage. This does not render the space without purpose, it is frequently used to mark special anniversaries or protests, as a highly internationalised space it provides visibility to a sometimes essentialised city. For example, in June 2014, when the Global Summit to End Sexual Violence in Conflict was taking place, supporters unfurled a banner over the Old Bridge (Fig. 6.7).

Fig. 6.4 'DON'T FORGET' rock Mostar (Photo taken by author April 2015)

Using a highly photographed tourist location to engage with a political summit demonstrates the capability of actors to rescript space, albeit temporally, and to use institutionally staged space for their own needs. This connects the international issue to the local setting, which in the context of Mostar and BiH, is particularly relevant as sexual violence towards women was a weapon of the 1992–1995 war with many victims still without access to justice and support in the post-peace agreement BiH (Amnesty International 2017).

Similar to spaces outlined in Chapter 1, the Old Bridge has an intense gravitational pull for social movement, while this may now be largely by tourists, the popularity of the narrative of the bridge has been established transgenerationally. While there is a conflict of narratives at the site of the bridge, it can be observed that local actors have accepted the international institutional staging of the bridge, and rescripted this with their own contributions to the narrative of the space. This does not make the space any more local than it is global, and such interpretations of space

Fig. 6.5 'DON'T FORGET '93' rock Mostar (Photo taken by author April 2015)

occur momentarily. The use of the bridge as a site to make protests visible, such as the commitment to 'End Sexual Violence Against Women in Conflict' demonstrates the intrinsic global-local nature of spaces and the ability of local actors to use the visibility of internationalised spaces, for local (and global) issues. This is not to overlook international generalisations about the bridge, but to demonstrate that these narratives may

Fig. 6.6   Plaque on a piece of the Ottoman Old Bridge (Photo taken by author April 2015)

only have an audience in the international. Of course, this is problematic, as frequently Western researchers (I am included in this category) set research agendas, and therefore have a great power in establishing narratives of space. However, local actors can and do use space to their own ends, and in ways which are not always visible. What this demonstrates, is ultimately the complexity and fluidity of narratives in space which may be unseen or only temporally engaged with, but this does not negate the transformative potential of movement in them.

The Partisan Memorial

The next space we come to is the Partisan memorial located in the west of the city. The popularity of the space can be observed in the frequency of which it was mentioned by participants. Though it did not feature on hand drawn maps, the Partisan memorial featured in most narratives as

Fig. 6.7 The Old Bridge with 'End Sexual Violence Against Women in Conflict' banner (Photo taken by author June 2014)

a location which should be renovated. Bogdan Bogdanović[3] designed the memorial which is now 'considered one of the country's greatest twentieth-century cultural sites and is a designated National Monument' (Walasek 2016: 91–92). The memorial also functions as a cemetery for partisans who fought fascist forces in Mostar in the 1930s, and preserved part of the city through preventing the 'the [planned] destruction of the bridges' (UNESCO 2005: 26). The space of the memorial is somewhat divisive, though it is regarded by some as a space that transcends ethno-nationalistic identities;

> [i]t is architecturally fantastic; the symbols have very deep meaning there are no flags there or pointed nationalism. It is popular with middle-aged Bosnians, everything there [at the memorial] has a meaning. (J Interview 2015)

Through the complex scripting and heritage of the memorial the space can be read as one which does not replicate the ethno-nationalistic divisions in the city. The appeal of the space to middle-aged Bosnians can be linked to those would have experienced life in Yugoslavia. Despite the status of the memorial as a designated National Monument in the post-conflict space, the memorial has been neglected and was until April 2018, in a state of disrepair. As observed in the report 'Commission for Reforming the City of Mostar' by Winterstein (2003: 65), local authorities considered destroying the memorial 'with the site planned instead for public parking'. As evidenced by Winterstein (2003: 65), and the historic state of the memorial, there has previously been a lack of institutional responsibility for the space (Ethnographic research, September 2015). While there have been social initiatives that aimed to combat the denegration of the memorial, without the applied institutional support and protection, the preservation of the memorial had been difficult to maintain.

As the memorial was mentioned in many informal conversations and interviews, if a participant did not mention the memorial as a space of movement, I would ask about movement to the memorial. As C had been in the city before, during and after the war, and spoke about their movement 'all over Mostar,' I asked C if they ever spent any time at the Partisan memorial; '[t]he Partisan Cemetery is in a very bad state, it is vandalised and full of litter; it is a shame.' Before the start of the interview, C had told me about some different social activities she organised for her students, one of the activities was a litter pick-up on the highway;

Me: Maybe you could do a litter picking / clean up exercise there?
C: I would not take students there to do a clean-up exercise though [laughs] that would be... [Participant cringes]. (Interview 2015)
Me: Would it be a political statement to do this?
C: [Nods, continues cringing].

C's narrative of movement, or lack of movement to the memorial demonstrates two core pieces of information about the memorial: the first is that it had been in a derelict state for a while, the second is that divisive political and social meanings imbue the space. The memorial, as previously mentioned, exists in a Croat city area, and has significant meaning, largely, for Bosniak populations in Mostar due to the divide of the city during the Second World War. The location of the memorial indicates that the city was not previously ethno-religiously divided to the extent that it currently is. It also demonstrates the persistence of conflict divisions from this time period, all be it, exacerbated, generalised, and emplaced on to the narrative of the 1992–1995 war.

There has been significant debate over the role of political actors in preserving (or not preserving) the memorial and the assumption of a general dislike of the memorial in the West of the city, according to K, 'the people in that part [of the city] don't really like it' (Interview 1 2015). The memorial is also a site of exchange of fascist and anti-fascist graffiti which is indicative of the conflict of symbols that take place in Mostar. What is visually identifiable is that the graffiti has been tagged on the memorial at different times, which indicates that both individuals with fascist, and anti-fascist ideology use the same space. The presence of the graffiti demonstrates that rescripting space is temporally dependent and that space has layers of identity that may not be visible or engaged in different spatio-temporal moments. Though J noted the ideological origins of the memorial, she underscored that graves in the cemetery are ethnically diverse and that 'regardless of names [those who fought and died] were primarily young' (Workshop 2015). In focus, for some, the use of the space is not directed by ethno-nationalistic divides, but ideological ones; '[t]here's an ideological dimension to those that actually use the space, even if it's in disrepair from dirt, trash or broken glass, people will still visit it' (J Workshop 2015). On the 14th of February 2015, the 60th anniversary of the liberation of Mostar from the German and Croatian military forces by the Partisan forces, the site of the memorial was blocked;

> '[t]here was a big fire there [on the path leading to the memorial] of piled tires to block access to the memorial. (J Workshop 2015)

The previous observations demonstrate that the ideological dispute over the space remains a source of conflict in the city. As can be seen above, and what echoes official reports, one of the main reasons that the memorial has previously been in a state of disrepair is the lack of institutional support for the maintenance and upkeep of the site. Furthermore, it is identifiable that for some individuals, ideological divisions mirror current ethno-nationalistic divisions. Notably, the interest in the memorial, highlighted by most participants, and the contestation over the space through graffiti demonstrates the transgenerational passing of conflict divisions stemming from the Second World War.

> There is the Partisan memorial, there should be flowing water there, there was a plan to reconstruct, but it is on the Croat side. I like them

[Partisans] some of my family were Partisans. They are really cool, they are in movies and kill like fifty Germans, in movies like The Battle of Neretva. (D Interview 2015)

D's brief account demonstrates the transgenerational narrative of heritage, the heritage of space, and of the Partisans to their identity as a young Bosniak in contemporary Mostar. This observation further demonstrates D's understanding of the complex divided nature of the city space of Mostar, noting an awareness of not only spatial, but the ideological divisions in the city. D's awareness of the spatial-temporal milieu of the physical staging of the memorial and social scripting of the meaning imbued in the space is also demonstrated through this narrative. However, it notably illustrates a romanticisation of the narrative of the Partisans and of war time violence. Palmberger (2016: 153–155) observed a similar nostalgia when attending a 2008 commemorative event for the memorial, alongside the scripting of the commemoration as being primarily directed towards Bosniaks. This is illustrated through Palmberger's observations regarding the sites visited ('Šehitluci...where most commemorations for Bosniak war victims take place') language and religious verses consulted, alongside symbolism such as the fleur-de-lis, which is associated with Bosniaks and the ABiH (2016: 155–156). Moreover, as part of the memorialisation the group visited Abrašević, Palmberger (2016: 153–154) provided an account on entering the main hall;

Tito songs blast out of the speakers, occasionally interrupted by short sequences of one of Tito's speeches. Obviously recognising the songs, the old people smile, clap or sing along. I realise that a woman sitting in the row before me has tears in her eyes. On one occasion, at the end of a sequence of Tito's speech where the audience on the recording applauds, people in the present audience applaud too.

In Palmberger's account the those delivering speeches at the event then moved on to note an on-going fight against fascism in the city, underlining the present threat Mostar faces but also the ability for BiH to 'overcome national divisions as it has done so once already in its history after WWII' (Palmberger 2016: 155). During the 2008 commemoration, the memorial was in a poor state, as Palmberger (2016: 151) observed while it was the architects intension 'to integrate the memorial into the landscape [...] now it seems as if the landscape is taking over the memorial'. From pictures Palmberger (2016: 152–153) provides of the memorial in 2008 it is clear the state of the memorial worsened over time. As Fig. 6.8 demonstrates in 2015 it was significantly overgrown; the memorial stones were hidden from view by the untamed foliage or had been broken (Ethnographic research, September 2015). The memorial had been staged in a derelict condition due to institutional neglect for a long time which produced local calls for the space to be socially rescripted. The social desire to transform the space was identifiable through the creation of a Facebook group 'Partizansko spomen-groblje – Help to preserve famous WW2 memorial in Mostar'. Additionally, the memorial was used as a location for the City of Lights festival organised by two Abrašević affiliated participants, H and G. The light festival was organised as part of a Youth Art Movement project jointly organised by the Youth Council of Mostar City, the UWC, and the NGO Youth Power. The mission of the NGO involves the;

Fig. 6.8   Partisan Memorial Mostar (Photo taken by author September 2015)

Promotion of healthy lifestyles, non-violent behaviour, gender equality, connecting young, active participation in social life and decision-making, volunteerism, promoting culture, advocating for democracy and human rights. (Youth Power 2018)

The City of Lights involved artists from Sarajevo, Opuzen and Berlin who used public spaces for light installations. The inclusion of the partisan memorial in the City of Lights Festival, encouraged movement to the memorial at night, temporally transforming the use of the space which is frequently avoided at night. These outlined initiatives represent the agentive capabilities of individuals to enact against the institutional staging of the space and to socially rescript space in the city. However, it is important to note that such rescripting has a temporal impact.

Preserving and Restoring the Partisan Memorial

The Partisan memorial was mentioned by several participants as space with the potential to be transformed, though, due to the position of the monument, some speculated that Croat political interests have prevented the restoration and preservation of the space;

> The Federal Agency for Protection of Monuments has jurisdiction over the Partisan memorial. And any commercial activity in that area. There is no public discourse about this. There is an initiative to preserve the partisan monument. People to raise funds, somebody cleans it. This is all symbolic activity. There is the Ustaše, Croatians, Croats, Officials and residents, then the communism and anti-fascism. The peoples liberation fight... Communist/Yugoslavians are automatically segregated as bad, [but] ideologically it appeals to young children and youth. (J Workshop 2015)

As can be seen here there is notably a complexity to the divisions at the site of the memorial. Participants' H and G both mentioned the preservation and restoration of the memorial, a perspective shared by many in Mostar, as evident through the online support for restoring the memorial. The previous institutional staging of the site as derelict was considered by numerous participants to be instrumental as previously noted according to K and D, the residents of the Croat side of the city do not like the memorial (Interview 1 2015; D and E Interview). Furthermore, the space of the memorial is notably a space of the continuation of the conflict, several participants observed a political willingness for the memorial to 'deteriorate as much as possible, then to transform it completely' (J Interview 1 2015). Due to the multiplicity of narratives surrounding the Partisan memorial it can be considered an important location in the city, as it represents an intersection between different divisions in the city. Spatially located in the West of the city, the memorial has become a contested site due to the ideological history of the site and the links to anti-fascism. Despite the contested narratives of the space, the memorial is popular space in the city for renovation, one that is considered to have an aesthetic, cultural, and transgenerational value beyond the ideological ties of the space. Physically the Partisan memorial is in a Croat city area, and for this reason, it is has previously fallen into disrepair due to the institutional neglect of the area. Regarding conceptual or figurative spaces, as previously mentioned, it can be surmised that the Partisan memorial, has a distinctly complex ideological and historical narrative. However, in 2018 the memorial was included in the 2018 city budget, and in early April reconstruction and restoration activities commenced at the memorial (City of Mostar Budget 2018; Šotrić 2018). By the end of the April 2018 the overgrown foliage had been cleared and a large amount of the graffiti had been cleaned from the memorial (Bljesak.info 2018). Additionally, a security guard hut with security lights was also installed at the entrance to the site, this has reportedly increased the movement of tourists to the memorial (Bljesak.info 2018). Fundamentally, this institutionally supported renovation quickly transformed the space and implemented safeguards in order to dissuade vandalism at the site. Such activities have the potential to establish a longevity to the restoration of the memorial, and also demonstrate the capabilities of institutional actors to support and create opportunities for spatial transformation, even in highly contested spaces.

THE KOSAČA CULTURE CENTRE

The Kosača culture centre is the next space of interest, first constructed as a cultural centre in 1960, it was reopened in 1994 as 'Croatian House Herceg Stjepan Kosača' named after a Croatian Duke (Kosača 2016). Notably, the centre has ethno-nationalist origins which prioritises Croatian cultural events. In 2003 (65) Norbert Winterstein, in the Commission for Reforming the City of Mostar report, observed that the use of the building is predominantly by the 'Mostar City-Municipalities with a Croat majority' and that the title of the centre and usage is 'considered by non-Croat citizens of Mostar as segregationist'. The report regarded the building as 'a blatant example of division', one which the city should seek to address (Winterstein 2003: 65). Over ten years later the centre remains in the city, it is a large building on one segment of the Rondo, established during the Austro-Hungarian time of development in the city. It has a wide forecourt, which has a large sculpture (described as 'a monument to the fallen Croatian defenders') when standing in front of the building the sculpture forms a crucifix through the layers of stone stacked to produce the image of a cross (Kosača 2016). An additional signifier of the space is the presence of the HVO military flag outside the cultural centre, this is the flag of the Croat military force, which was in operation during the Bosnian war. The presence of the Croatian flag, the HVO flag, and the crucifix sculpture are visible in Fig. 6.9.

Fig. 6.9 The Kosača culture centre (Photo taken by author in April 2015)

The cultural centre was a frequently discussed space in the city, sometimes as a point of the division regarding the events which the centre promotes and puts on. According to participant J;

> the primary point of Kosača [was] to serve the Croat national assembly in the war, and it remains so, divisions remain in urbanistic ways, there's a change of play into Croat divisions in the hope of a seat of Croat power, of establishing Mostar as the Croat capital of Bosnia. (Workshop 2015)

Notably for J the centre is a divisive space which maintains the attempted staging of Mostar as a 'Croat seat of power', this echoes the historical claims once made by Croatian President Franjo Tuđjman (Bollens 2007: 170). This staging was also evident to participant B, who reflected on folk music and the dynamics of nationalism inherent in this type of music. It can be regarded that folk music is tied to identity and heritage and evokes the cultural differences that become a signifier of the divisions of the conflict (Kaneva 2011: 215). According to participant B the centre lacks appeal as it only organises concerts for 'Croatian musicians. Like Klapa, seven men who sing together some special sound singers from Croatia' (This is an acapella band from Croatia). For B (Interview 2015) Kosača, which operates as a music venue and cultural centre, 'could be more international.' I had taken the picture which is Fig. 6.9 the day before the interview, on the 8th of April, so I asked B about the presence of the HVO Croatian military flag. B shrugged and said that it was 'some national day of Croatia' (Interview 2015). The 8th of April was the day of the creation of the Croatian Defence Council (HVO) which preceded the 1992–1995 Bosnian war (Solis 2016: 727). Notably, through B's phrasing of the day as '*some* national day of Croatia' [emphasis added] the wording can be interpreted as indicating a sense of detachment, reflecting a level of disinterest with nationalist discourse. The basis of the statement, however, does indicate a level of awareness regarding the raising of the flag at the cultural centre.

Institutionally, the presence of the flags are a visual signifier that mark the ethnonationalistic ownership of space. The raising of a military flag outside the cultural centre in Mostar ties the narrative of Croatian militarism to the space, which is utilised largely as a cultural venue. In this respect, the presence of the flag represents the conceptualisation of the space, by top-down actors and stages the space as divisional. However, there are multiple variables to the use of space and it is the tendency of the venue to invite only 'Croatian singers', and the folk/acapella style of the music playing at the venue, that influences B's avoidance of the centre. The variable of music impacting on movement, represents a layering of interests in how the space is interpreted by the individual. In many narratives the HVO flag may be a distinct marking of territory, however, it is important to reflect on participant B's self-disclosed national identity as Croatian. Fundamentally, nationalistic symbolism may not deter movement if it does not represent that the individual is "out of place". The nationalistic dimensions of the centre were also discussed by other participants. C, who is a Croatian citizen, reflected on their movement to the centre, and noted that the centre is involved in a wide range of activities such as the Blues Festival and UWC graduation;

I like Kosača, [...] it is the national theatre—they have the Blues Festival there...we had the graduation for the UWC students there. (Interview 2015)

The Blues Festival is organised by the Rock School and is also supported Abrašević. The use of the Kosača centre as a location for The Blues Festival in 2014 (organised by the Rock School and supported by Abrašević) and for the graduation of United World College (UWC) students, rescripts the space as one that engages within and across divides in the city. The involvement of the Kosača centre with the UWC links the centre to the city-wide network of actors, and locations involved in facilitating spaces that promote conflict transformation in the city. While for some the cultural centre would not be considered the "national" theatre, through involvement with the UWC and the Rock School the cultural centre links with shared spaces within the wider city.

Fundamentally, the staging of the cultural centre is notably divisive, in particular the raising of HVO flag (alongside the Croatian state flag) militarises the already ethnically divisive space. Despite this, for some research participants, the institutional staging of the space as a location of Croatian national identity, is not part of the narrative of their movement or non-movement. This illustrates the layered experiences of space, which are only visible through reflective narratives of the use of space. Furthermore, for C, Kosača is not a nationalised space as it is used for non-ethnically divisive events such as the UWC graduation and the Blues Festival. This demonstrates the ability of social actors to rescript space for their own purposes. The extent to which this transforms the space for other actors in the city is hard to discern. However, over time, through events which are shared or more broadly inclusive, the narrative of the space may be transformed. Through diverse events held in the space, Kosača can be considered rescripted for a use that transcends the staging of the physical space. This rescripting, which may be temporal, evidences the agentive capabilities of social actors to transform ethno-nationalistically divisive space in subtle and perhaps time and context-dependent ways. Notably, this is instigated and supported by some social actors in the city, though this has not translated into long term restaging of the space.

The Old Glass Bank—Staklena Banka

It is used as a ruin, why use it as a ruin? (J workshop 2015)

Though the bank has been derelict for more than 20 years following the end of the conflict, the view of the city afforded to those who climb to the top of the bank makes the building a popular social location. Structurally, the building evokes the memory of the conflict as one of the many severely damaged buildings, but it is also a point of social rescripting, in that it is a prime location for street art produced in the Street Arts Festival in Mostar, as demonstrated in Chapter 5. The use of the bank as a space of art can be observed as important, due to not only the availability, but also the centrality of the space. The use of the derelict space as a location for public art, rescripts the space and with numerous pieces of art visible through their placement on the periphery of the Old Glass Bank, the building is transformed in its function as the space becomes an interactive gallery (Figs. 6.10, 6.11).

While building itself (and the bottom of the building, in particular) has been used by artists during the Street Arts Festival, it is also the location of some non-street art-related graffiti such as the words 'DIVIDE AND CONQUER' (written in BCS pictured in Fig. 6.10) which appears to be a commentary on the function of the political divides. Despite it being derelict, The Old Glass Bank is also a social space for some youth in Mostar;

Here, the bank…is in ruins, but there is a good view from the abandoned bank, it gets so crowded on the roof. (B Interview 2015)

Fig. 6.10 The Old Glass Bank (Photo taken by author in September 2015)

Through this, B notes the view from the top of the bank as directing their movement and that of other young people to the space. This narrative of movement reflects the use of the city space by young people who rescript the derelict space, as it is free to access, while affording them privacy, alongside a view of the city. It is observable that for young people, the popularity of the top of the derelict building, as a space to congregate, reflects a desire to view and not be viewed, and reflects a duality of being in public space, while not being seen (by some) in public space. The space is central but out of the way and though the building is abandoned, with movement discouraged, young people who want to climb to the top find a way. As participant F noted, when they had arrived in Mostar there was a gate in place to prevent access; but that nevertheless 'a lot of people spend time there' (Interview 2015). The Old Glass Bank reflects the way public space can be physically staged for one purpose but socially scripted as a different space, which generates an alternative platform for engagement with the space. In this example,

Fig. 6.11   ZAVADI PA VLADAJ (DIVIDE AND CONQUER) (Photo taken by author in September 2015)

the social usage of the space transcends the staging of the space as derelict and provides an alternative function of the space. Therefore, it can be observed that social movement transforms the space from a derelict building which maintains a loss of public space due to the conflict, into one that is used as a social space. The loss of public space in Mostar has taken two forms. Firstly, the institutional staging of the conflict divided the once unified public space of the city, resulting in a loss of public space for some through divided city services. Secondly, during the conflict, buildings were severely damaged in the city as noted in Chapter 4, with 'between 60 and 75 percent of buildings [...] destroyed or severely damaged' in Bosniak areas, while an estimated 20% were destroyed on the West side with this also mainly occurring on the frontline (Bollens 2007: 171). Though imperfect, and unsafe through social usage the site is rescripted as a social space for groups of youths drawn to the space for the view, and the privacy of the space. As part of the research, all participants were asked about the potential future use of the city space in Mostar, several participants noted the Old Glass Bank as a space that, while popular socially, needed renovation or transformation;

> D: The Old Bank is here, but I am forbidden to go to the Old Bank. I don't like to go anyway, there are needles, etc.
>
> E: I went once, it's dirty. A lot of people go from World College and the Gymnasium, a lot of people go. Usually twenty-ish [sic], go to Old Bank, go to the top, the view is the best view. It's hard to go there so... it is a dirty place the Old Bank. (D and E interview 2015)

Both D and E highlighted the bank as a potential space for investment in the city, due to the poor physical condition of the building, juxtaposed with the relative popularity of the space. While for participant B, the poor condition of the building did not deter movement to the space, both D and E observed that this influenced their avoidance of the space. Furthermore, for D, their movement is directed by a figure of authority in their life who forbids movement to the derelict building. Though E highlighted past movement to the bank, they noted that, despite the reported view, this was not an experience they enjoyed, due to the derelict condition of the building. Furthermore, participant E noted that the top of the building is hard to get to, this is also inferable from the participant's perception of the space as one that is dirty and unclean. Comparatively, an ease of movement to the bank is observable through participant B's narrative, as 'it gets so crowded on the roof' (Interview 2015). Through these differing perceptions of the space, it can be discerned that movement and use of space is influenced by personal perceptions, and immediate social circles who influence and direct movement and can also affect sensations and emotions produced by the staging of the space. Fundamentally, through movement some social actors have scripted the abandoned building as a social space, however, the scripting is limited and not supported by institutional actors. The maintenance of the staging of the building in a post-war derelict condition evidences the lack of institutional support or ability to transform the space. Overarchingly, this influences movement to, and use of the space, as outlined by participants' D and E, the space does not appeal to all. Fundamentally, the various narratives of the use of the space of the Old Glass Bank demonstrate that multiple social rescriptings' can take place in one staged location. From this, it can be proposed that social scripting can correlate or clash with institutional staging, and due to this flexibility, it can replicate the staging but also has the potential to produce new meanings in space, beyond those which are institutionally staged.

The Old Glass Bank is on a former frontline of the conflict and remains derelict while most of the area around the bank has been transformed; for example, the Spanish Square, the Gymnasium Mostar, and nearby spaces such as Kosača, and OKC Abrašević. The Old Glass Bank was once considered a potential site for the Abrašević youth centre but there were significant problems with the property rights of the building (H Interview 2015). Despite safety concerns, the space remains popular either for the roof or to see the art. Fundamentally, the draw of the space has untapped potential, as G noted 'renovation of the Old Glass Bank would be great', adding that a particularly important use of the space would be 'some kind of central youth culture zone' (Interview 2015). However, in February 2016, institutional plans for restaging the bank into a government building were shared with citizens in BiH and as of April 2018 this process is underway (Komora 2016; Street Arts Festival 2018). This renovation is significant, as it will transform the city by removing one of the most prominent spaces for street art. While it will take away a central space for youth to congregate in the city the space, in this context, it was not fit for purpose. The space is not safe or clean, empty lift shafts, exposed edges, broken glasses and syringes are some of the dangers of the space. This has been underlined by a recent deadly accident when a young man fell from the bank, which has reaffirmed the importance safe social space for youth in Mostar and has led to the Youth Council of the City of Mostar calling for all possible routes into the bank to be sealed (Youth Council of the City of Mostar 2018). While long term, the planned renovation will restage the abandoned derelict and dangerous building, alongside the need this demonstrates for more safe social spaces for youth, this will also result in the loss of a socially scripted space for street art in the city.

OKC Abrašević

Moving further north in the city and two streets East, we come to the OKC Abrašević a cultural centre, which is a self-styled shared space in the city. Due to the purposeful creation of the centre as a shared space for all, Abrašević can be considered a key space in the city. The origins of the centre as it stands today begin in the post-conflict city;

We were young and crazy and we wanted to build a new world, non-divisional and anti-fascism. At the end of the war there was a project ran by the OSRC [Organisation for Security and Co-operation in Europe]– mobile cultural centres, for three years they travelled through cities in FRY. Over there [one street over West from Abrašević] they settled them [shipping containers] there. (H Interview 2015)

For some time, the centre was a set of shipping containers, which are still used by the centre for offices. According to participant J, who was also a central actor in the establishment of the centre; 'Abrašević worked as a zone without the building containers, it moved here (to the current location) and worked in two places, we came from all sides' (J interview 3 2015). J's narrative of the establishment of the centre demonstrates that ethno-nationalistically mixed social actors facilitated the centre and reflects what H noted as the non-divisional aims of the centre. During this time, those working on the centre were between the 'two spaces' they had to move between 'Abrašević and the mobile cultural containers' which was tiring for H 'because you were never out of the space' (H Interview 2015). While other spaces, such as the Old Glass Bank were considered as spaces for the establishment of the Abrašević centre, the current site was chosen due to transgenerational relevance as it was the site of an old socialist workers club. This was a popular space in the city according to H, who observed that it 'was important before the war, [as] many people met [at the centre] and there was a positive effect in public space' (H Interview 2015). Following an issue with the ownership of the building, 'on the 20th of August 2004', the centre was finally registered (H Interview 2015). After this success, H recalled that the 'OHR wanted to represent [the centre] as a "moment for unification" this was a problem' (Interview 2015). As H outlined, the process of the establishment of the space was not without challenges. At the time of the OHR wanting to represent OKC Abrašević as a '"moment for unification"', those involved in the centre included 'a strong group of anarchists,' who contested the involvement of the OHR with the centre (H Interview 2015). In the interview, H a former director of the centre, reflected on this internal conflict and surmised that 'I decided I will go in the direction of diplomacy, I stood on the side of allowing him to come (to the centre). Some left, but they came back' (H Interview 2015). The narrative of this disagreement demonstrates the contentious involvement of international actors in the establishment, or endorsement of Abrašević.

This element of negotiating ownership and support is demonstrated further as H elaborates; '[w]e were offered American money, but we refused the first time as we had criticisms of America regarding the war in Iraq' (H Interview 2015). However, issues with the construction of the building led to the acceptance of the funding for building a flat roof that would allow for future development of the centre (H Interview 2015). During construction and reconstruction, as J outlined, those involved in establishing the centre 'worked [and] were open throughout all these situations and conditions' (Interview 2 2015). Following a long process of ensuring ownership of the space, the Abrašević team legally secured the current location for the youth centre. The opening of the centre was around the same time in 2004 as the completion of the reconstruction of the Old Bridge. With the Old Bridge regarded as a unifying structure, the opening of Abrašević, though technically a unifying space, passed largely unnoticed by international organisations. This is, in part, due to the topographical significance of the Old Bridge to the city of Mostar. However, Abrašević, importantly facilitates a space for the thoughts and actions of re-unification of the city.

ESTABLISHING OKC ABRAŠEVIĆ

The work of socially establishing and maintaining the centre physically and conceptually, while internationally and nationally supported by institutions and funding bodies, was locally driven. Alongside, outlining the aims of the space to transform the divided city, participant H noted the how they interacted with the space, '[f]or me, I'm still a member of Abrašević, but I'm not active anymore, I do a little' (Interview 2015). H elaborated that they stopped being an active member in 2005 but;

> [i]n 2012, I came back here to work; it was hard for me to come back and invest a lot of energy and do a lot of work. Maybe I did expect support from the people, but especially this year I see the support that people appreciate this work. (Interview 2015)

Through H's description of involvement with the centre the emotional energy that establishing a shared space requires from individuals involved in the post-conflict space is evident. In discussing the time of labour, and what H would regard as the time when they felt appreciation and support (the interview was conducted in 2015), it has taken years of labour, without considerable personal gratification from social affirmation. In this, it is important to reflect on not only the largely unpaid nature of such locally established shared spaces but also the sometimes emotionally unrewarding nature of such work. For H, there was no timely appreciation for their involvement in establishing Abrašević. This personal relationship is often overlooked when researching post-conflict spaces and reflects the devaluing of the personal and emotional, which is critically framed as the feminine. This overarchingly mirrors the trajectory of international relations and post-conflict peace processes as the local and, in focus, the private mechanisms of peacebuilding are sometimes overlooked. While the goals of establishing the centre were to create a space of peace, the presence of conflict is identifiable in the process of the establishment of shared spaces. Additionally, H considered the significance of the centre on their individual movement and agency '[f]or me, Abrašević is like home it's touched me a lot, but I want to do my own stuff, to have my own agency' (Interview 2015). In this statement, H observed the intrinsic importance of the centre to their identity, but also noted that it has limited their personal interactions (H Interview 2015). For H, this translated into a limited spatial landscape which curtailed their use of the city space;

> I love Mostar, but it is a pain in the ass, it's hard with the people, last year I opened myself and went to more places, when I worked in Abrašević I was just here. I went back to the street and opened myself [to different places in the city]. (H Interview 2015)

For H, though Abrašević is undoubtedly an important shared space in the city of Mostar and one that arguably exists, in part, due to their own hard work, it became a limiting space for her movement. Critically, space can become conceptually closed even when it is shared and, in actively working to create a shared space, any movement, even movement within shared spaces or spaces of peace, can become socially divisive. Through this the complexity of space and interactions therein can be observed. Even movement to shared spaces, due to the social scripting of the space, can entrench movement along established routes and can become socially limiting, despite the best intentions of those using the spaces. What can be taken away from this is that the establishment of spaces involves an interactivity of social and institutional actors. Furthermore, the example of Abrašević sets out the capability for social actors to rescript meaning in space, while also underlining the vulnerability of individuals who seek to foster shared space in post-conflict cities.

THE SPACE OF OKC ABRAŠEVIĆ

Abrašević is open to all ages and hosts a variety of events with a wide ranging, and notably a transgenerational appeal. The centre requires no talent or educational ability to utilise the space and individuals can attend performances and talks. The widespread inclusion of Abrašević in participant narratives evidences the centre as an important or fun space in the city of Mostar. In particular, participants consider the centre to be a desirable location for music and coffee, 'a good place to chill out' (A Interview 2015) and 'a benefit [to the city]' (F Interview 2015). The popularity of the centre reflects the social importance of the space. While one of the aims of the space is to traverse institutional divides and numerous participants highlighted the centre as a location to relax and to spend time in the city, no participant specifically regarded the space as a space for transcending the divisions which persist in the city. This does not mean that the space is not effective in this but to highlight the dual use of the space that enacts one purpose (cafe, concert venue, socialising space) and conceptually performs another function as a space of peace. The space of Abrašević seeks to traverse divides of the conflict through "neutral[4]" activities, 'to be able to bring closer the youth of so-called, both sides of Mostar' (J Interview 2 2015). Accordingly, the centre operates without any nationalistic agenda, and the openness of the space is noted by Kappler (2014: 174) and Carabelli (2018: 129) who also observe that the Abrašević centre has extensive connections in the community. A large and somewhat fluid network of social actors (as members or associates) established and maintains the centre, and through founding Abrašević a human network of cooperation was fostered, one which traverses the institutional and social divide of the city.

However, no space is entirely open, as social meaning is attributed to space. While the centre operates as one that is open, it is also closed to some at the same time. In order to understand power dynamics of space it is important to understand how 'common space is established' Kappler (2014: 37). In focus, this is not to overlook the work of the centre; but to highlight the coded nature of space, which becomes scripted through specific social performances, this in turn, scripts the use of the space. Therefore, not only does the scripting of space impact on movement, but it shapes future scripting and directs social performance in space. It also guides social usage. Therefore, the Abrašević centre, as a space that actively seeks to traverse divides, is more likely to attract individuals that have an affinity with this goal. Furthermore, due to the ideological history of the centre it can be also considered that there is a latent ideological divide in the space which may detract some movement (Palmberger 2016: 160).

'Affinity Space'

While the shaping of the centre by local actors creates essentially a captive audience, this can mean than some are excluded from the space. As noted, though the space of Abrašević is open to all, those who utilise the space script the identity of the centre, therefore like all spaces in diverse places, not all residents in the city will feel comfortable going to the centre. Critically, we foster patterns of spatial movement to spaces that feel familiar or in line with the social identity we believe we have, or sometimes want to project. Fundamentally, non-movement to shared space does not necessarily mean that individuals are resistant to peacebuilding or shared spaces but may be representative of an 'affinity space', which Gee conceptualises as being borne out of eleven features (2005: 225). Gee (2005: 225) lists, that in such spaces, a 'common endeavour' takes precedence over other variables of identity such as race, class, disability, gender or sexuality. While affinity spaces as set out by Gee (2004: 85–87; 2005: 225) are formed of eleven principles, these cannot all be readily applied to Abrašević. However, the collaborative and interest driven nature of affinity spaces which Gee conceptualises is evident in the space of Abrašević and the concept is useful for outlining the draw of the centre (2005: 225). The space is shared in terms of the intended use of the space by residents from 'both sides' of the city but also the ability to engage with and influence the use of the space. This is observable in the narrative of the space that Abrašević puts forward through its physical and conceptual network connections in the city. However, as noted due to social scripting, all social spaces exist 'on a spectrum somewhere between being open and closed' (Kappler 2014: 28). Fluidity in the use of the space is something that for Kappler (2014: 28) indicates the extent to which the space is 'open'. Critically, this does not always mean that space is open or always shared, and barriers to accessing the space may exist in the form of 'physical or ideational distance, processes of "othering", the use of coded language or jargon within a space and so forth' (Kappler 2014: 28). As previously identified, while the space of Abrašević is conceptually open to all, it can be regarded as closed to some, due in part to the ideological heritage, which may be hard to disentangle from the scripting of the centre as a bohemian or 'alternative' space (Palmberger 2016: 160). Furthermore, the social scripting of the centre as one which seeks to traverse divides may disparage movement to the space, as the act of moving to the centre becomes bound up in addressing the discourse of the conflict. In a city with economic and employment issues, the likelihood of individuals having free time and emotional energy to dedicate to such activities is low (Bramsen and Poder 2018).

Abrašević is notably both a space and a network of actors, it occupies the physical space of the centre, and creates a conceptual space for the active mixing of the citizens of Mostar. Through this the centre facilitates the rescripting and growth of other spaces in the city, as can be seen through the Street Arts Festival supported by Abrašević members. The network of members' of Abrašević continue to create an inclusive space in the city for all ages, and groups. The centre is therefore pivotal as a shared space which facilitates wider engagement in the city. This engagement is directed by the establishment of a physical space, and the local social actors who direct the rescripting of the centre. Theoretically, Abrašević acts as a nexus due to projects stemming from the centre itself, and the network which supports the centre in the city. Notably, there is no comparative space to Abrašević, it is a unique space in the city physically, and conceptually, through the programme of events organised at the centre and its outward, flexible, and adaptable outlook. It exists uniquely as a space open for anyone who wants to attend, and one that is free of nationalistic symbolism (Kappler 2014: 173–174; Carabelli 2018: 129). Fundamentally, the space of Abrašević is reactive to the city, and the purpose of the space, or what the space provides, can change with the needs of the city which will inevitably develop as the community changes through time. It can be regarded as a mutually sustaining relationship as local actor's interests shapes the activities the centre establishes.

Fundamentally, OKC Abrašević is conceptual and physical; it is a place that has been physically built, but conceptually the space is constantly transforming. The space exists liminally between the physical construction of the centre, and the social construction of the purpose, meaning, and use of the space of the centre. The success of the centre represents a merging of local and international actors which is emotively driven with many local people working at the centre without a full salary, performing emotionally draining work (J Interview 2015; H Interview 2015). The social scripting of the centre is the foundation of the network which Abrašević fosters. In summation, the space of Abrašević, though notably assisted institutionally in its staging, has been significantly supported at a social level. Fundamentally, social scripting, has supported the establishment of the Centre, through a combination of volunteer work and paid work. The local social involvement in the space has sustained the centre and fostered a network stemming from the centre. Ultimately, the space of Abrašević demonstrates the pivotal role that social actors play in transforming post-conflict space and fulfilling the aims of a shared space.

MEPAS MALL

Moving North and several streets West in the city, the next space we come to is Mepas Mall. The mall was opened on the 12th of April 2012 and is the busiest shopping centre in the city of Mostar, with most commercial units occupied and open for business (Ethnographic research April 2015, September 2015). The mall exists in a Croat city area though the space is largely internationalised through the companies operating at the site. As one of the newest and largest buildings in the city of Mostar, Mepas Mall is visible across the city. The mall has an on-site security team and a five-star hotel which is on the top floor of the shopping centre. It is a privately-owned space but one that can be considered shared in that it is open to all individuals for their usage. The mall attracts a large footfall and the cafes in the mall are visibly as popular as the cafes in other areas of the city (and when the weather is poor-more popular). Additionally, there are unique shops and spaces to Mostar in the mall, including a fast food restaurant McDonalds, multiple international clothing stores, a cinema, and bowling alley. Notably, some participants noted that in the cinema, films subtitles are in Croatian. While Bosnian and Croatian are very similar they have minor differences. The use of Croatian subtitles parallels the ownership of the space by a Croatian company. Despite this, the space is not considered divisional and the mall has a wide appeal, it operates as a place to meet, to eat, to be entertained, for visitors to stay, to shop and buy clothes and groceries. In the physical sense, Mepas exists in the local space, while conceptually or symbolically it is international. The mall and the type of consumerism it promotes is unique in the history of the city. There are other shopping malls in the city but none as encompassing for activities as Mepas. Previously during the socialist era, there was a shopping mall in the city called Hit, which was opposite the Gymnasium in Mostar;

> Hit was part of the narrative of the city, it was a department store place to buy your first record. Shopping culture, shopping has changed, today the cultural concept of shopping has changed. There was not so much choice, it was logical to buy everything there, first floor: school supplies, second floor, jackets and coats. Everything was uniform, plates et cetera. Uniforms, there were not many choices. (J Interview 2015)

For H also, the mall was bound up in nostalgia, with the participant suggesting artists recreate the façade of the building 'as an artistic movement' (Interview 2015). Comparative to Mepas, the space in which Hit mall stood, was historically, and would also in contemporary Mostar be, more central to the main transport and footfall arteries of the city which made the mall a central hub for social movement in the city. Nevertheless, Mepas remains a popular space in the city. What can also be noted is that, the space of Mepas is symbolic of the economic transition of the city and country at large into a capitalist economy. This has an impact on the type of consumerism alongside the interactions with and within spaces of consumption. The internationalised[5] space this creates, can be differentiated from other spaces in Mostar. Additionally, as a space of consumption, the mall can be considered inclusive to a point, as it is inclusive for all those who have money to spend in the space. This is set out by staging of the space, including advertisements and posters which, Benjamin (1986: 135) notes historically, as signs and symbols which set out the ideal lifestyle of the individuals who use the space. Notably, the social membership for interacting in, and moving to, the space of the mall is outside of any specific social identity or social capital. Though, for some participants, it is considered a negative space of consumerism. However, the space is important for many citizens, as noted by research participants, it provides a space for shade in the summer and a place for heat in the winter (Workshop 2015).

For numerous participants the space of Mepas, was highlighted as a functional space, 'I don't go to Mepas unless it's for Kozum (food shop) cinema or to pay my mobile bill' (C Interview 2015). In this, the participant distanced themselves from the space of the shopping mall to the exception of certain shops that exist only at that location noting that '[i]t is convenient you can park under it and do everything. But I'm not a shopping person' (Interview 2015). C further reflected on their movement to Mepas Mall, noting the only other time they would use the space socially is to take a young family member;

> when my nephew comes to visit, I take him to McDonald's, he feels in control in McDonald's because it is predicable, he knows exactly what to expect. Or I might take him to the Old Town for ćevapi, but with that, it is funny because you can tell where someone comes from by how they eat it [ćevapi]. (Interview 2015)

Through this C presents not only how other social actors (her nephew) influence their own movement in the city, but also how space and identity, and the staging of space, interplay with the movement and use of space. In focus, because the space of McDonald's is staged in a standardised way (Ritzer 2012), it is familiar, and C's nephew feels comfortable. Comparatively, he feels out of place eating ćevapi in the Old Town because the social norms or social rules of eating ćevapi (a meat kebab and flat bread dish) are different in that space to how he eats it. According to C, her nephew uses a knife and fork to eat the meal but Mostarians 'eat [ćevapi] like a sandwich' (Interview 2015). This indicates the scripting of actions to physical space, demonstrating how social actions and space inter-relate, in so far that some actions can be spatially dependent. Goffman (1971) identified this socially with regard to how we perform around others and that the outward performance may not be genuine, but it is what we wish to convince others of ourselves. While we may aim to tailor our performance in order to give off the impression we belong in a particular space, sometimes we may not have all of the necessary information in order to do so, or our learned behaviour can "give-away" our performance, as demonstrated through this narrative. This is an example of how small social cues can signal our understanding of, or establish our previous use of a space, and demonstrates the impact when these cues do not align to the space and therefore indicate a spatial otherness.

In general, Mepas is not a space imbued with symbolic or cultural meaning, it is simply a space to get coffee or shop and there are few emotions tied to the space of the mall. However, the lack of ethnonationlistic staging of the space, has unintentional potential to be functional in transforming the city space. A's narrative of their movement reflects the neutrality of the space, as they noted 'sometimes I go to the cinema at Mepas with my friends and we drink coffee where we can [with childhood friends] we are always interested in drinking coffee' (Interview 2015). For many participants, the mall is a non-emotive space, and it serves as a functional location for movement;

> E: Here is Mepas, I go there to go to the cinema.
> D: [Nods in agreement] Or I have to go with my girlfriend if she has to buy something. (Interview 2015)
> J: Mepas is only a strict destination, not a place you would hang around, not something on your walking route. (Interview 2015)

Though participant feelings about the space were not always neutral as G surmised, 'I don't like Mepas, Mordor... here is Mordor Mepas' (Interview 2015). Mordor is a fictional space that was created by J.R.R Tolkein, it can be surmised as a dark, evil place. Participant G's use of Mordor as a metaphor for the mall, therefore reinforces her dislike of the space. Similarly, participant F, though a regular vistor to the mall described it as 'horrible' (Workshop 2015). While J noted that from their 'personal perspective, [there was] no appeal' to Mepas as a space in the city, she still used the space from time to time including going there on a Sunday 'when everything else is closed' and for paying her phone bill (Interview 2 2015). Notably, feelings around use of the mall range from neutral or functional to inherently negative. It cannot be analysed as being an appealing space, it is for many simply functional space, even if they do not particularly like it. However, the variable of functional movement makes Mepas an important location in the city which facilitates social movement, from old and young Mostarians, different social groups, and ethno-nationalistic backgrounds. Fundamentally, the mall exists as a space, free of ethno-religious narratives and symbolism, as some participants noted, the mall operates as a shared space, in that it would not be unusual to see 'a woman in a hijab' in Mepas (in the example given by participant J). According to participant J, this 'familiarises people with the "other"' (Interview 2 2015). In subsequent questions, J expanded that;

> I think Mepas has achieved this position of a neutral place that appeals to people. They feel free to come there even wearing a hijab and not feel like they have transgressed into somebody's territory. (Interview 2 2015)

It is the somewhat internationalised space of the mall that provides a neutral ground for interacting with 'the other'; though the mall itself exists in a Croat city area. J categorised the space as one for 'reconciliation' (Interview 2 2015) as it functions as a positive middle ground or a neutral space in the city;

> [f]or people who have never seen this [the other] in their own neighbourhood it is a good space to see people and to be not so shocked to see the other. It's good to see the other. (Workshop 2015)

This above dialogue regarding the space of Mepas in the city characterises the overarching persistence of the divide in the city, but outlines the potential for the mall as a unintentional space of peace. As Cresswell (1996: 23) observes, there is a blurred line between transgression and resistence as spatially situated actions, in so far as 'the intention of actors' is not important in transgression, however the 'results – on the "being noticed" of a particular action' matter. Fundamentally, in this capacity the mall has the potential to facilitate 'everyday conflict resolution' (Mac Ginty 2013: 387) regardless of the intent of the movement, it is the act of being observed that instigates the spatial rescripting (similar to the dynamics of space-time interactivity). It is through the unintentionality of such movement that may include divisive actors who are marginalised (through their own movement) from shared spaces or spaces which facilitate transformation. Through this unintentionality, the space as one which facilitates social movement has significant potential in challenging conflict cleavages and social fissures.

Transgenerational Mepas

As discussed, the mall has a wide range of functional appeal. This is well illustrated by the transgenerational appeal of the mall, as observed by C (Interview 2015) 'elderly people go there in the winter to have a coffee and benefit from the heating' and also by J (Interview 2015) 'old people go, not so much linked to consumerism, there is a children's playground, free parking for 1 hour, and it is a meeting place.' The functionality of the space as a refuge in harsh weather was not simply related to the elderly as J surmised the air conditioning benefitted others in the city also, 'when it is 40 (celsius)...and you can't leave the house until 7 pm it is nice to go there' (Interview 2015). Through these reflections on movement, Mepas Mall can be observed as providing a social space that is unseen in its function. Furthermore, generational differences impact upon the reasons for movement. In focus, this provides a snapshot of conditions faced by the elderly in Mostar. As such, it is important to reflect on generational economic issues faced by residents as they cannot afford adequate heating and many more cannot afford air conditioning in their homes. As a space of modern consumer capitalism, the mall has a number of access points, and therefore provides an intersecting space of engagement across multiple divides, operating as an open space for many.

THE PRICE OF THE SPACE

Notably, the mall dissuades some individuals from using the space, as observed during field research, Roma who were forced to beg and had approached customers in the mall were quickly escorted to the exits (Ethnographic research, 2015). This is not exclusive behavior to the mall and Roma were frequently escorted or chased from cafes, shops, and bars across the city (Ethnographic research, 2015). Massey (1991) provides an important contextualisation regarding the way in which space, place and movement can implicitly "other" some persons. In a reflection on global flows of movement (Massey 1991: 25) notes that 'different social groups…are placed in very distinct ways in relation to these flows and interactions', this can be zoomed into the city level. The interconnection of globalisation and the divided city can be identified as what increases movement for some restricts movement for others, as Massey notes (1991: 26) 'some are effectively imprisoned' by global flows of capital. Fundamentally, this can be linked to the mall through the juxtaposition of the shared space it provides for some, and the barriers to access it creates for others.

As previously discussed, economic resources are an important part of the ability to engage with commercialised public space. This may only be the price of a cup of coffee, but the price of the cinema, shops, McDonald's, bowling alley are costlier than coffee. Moreover, according to J (Interview 2 2015) '[n]ot everyone uses it [the shops], even in a sale.' The lack of use reflects the unaffordability of the space of Mepas which produces alternative modes of engagement with the space, or low-cost modes of engagement. The mall, until early 2016[6] had been the location for a group of women from different ethnic backgrounds to meet, organise a weekly charity clothing auction, and exchange the auctioned clothing (J Interview 2 2016). The establishment of 'Humanitarian Auctions', in the space of the mall, rescripted the type of consumption. This initiative created the opportunity for local women to auction off old clothes and to bid for other items of clothing. 'You could get a new pair of pants,' J explained, and added that 'each week there is a new collection then they auction stuff, then whatever [money] is collected goes to the person they are collecting for' (Interview 2 2015). J was an active member of the group and went on to discuss how the group functions beyond narratives of ethno-nationalist divides;

> [w]e always meet in Mepas and it lasts, there's usually no perseverance, but this group works really well, it does it regardless of nationality and also through this [Facebook] group there are fifty active members, you will find people make friends it doesn't matter if their name is Alma/Marija. (Interview 2 2015)

Through this J notes that it does not matter if the women have Turkish or Croatian origins to their names, which would associate respectively with the Bosniak and Croat ethno-nationalistic groups in Mostar. The auction group itself exists across multiple spaces, existing online through the social media platform of Facebook, physically when the group meets in Mepas, and conceptually due to the diversity of the group as open to all. Through this, the group establishes a space of engagement which traverses the ethno-nationalistic divide of the city. The group facilitates an informal network of exchange, socially and monetarily through the humanitarian efforts to raise money for local causes, this sometimes involves funding a medical procedure for a resident (F Interview 2015). Through the functionality of meeting at the mall to informally exchange goods and money, the ethno-nationally diverse group of women socially rescript the staging and mode of consumption in the space. What is notable about this type of rescripting is that it is temporally dependent and facilitated by the physical presence of those organising the auction. Fundamentally, in what can be considered a public space, (though privately owned) the women who run the group created an alternative space of engagement which transformed the mode of consumerism at the mall, from a profit driven exchange, to one that is established to fundraise for those in need in the city. What is important about physical spaces such as Mepas, especially in a divided city, is not always their physical location, but how they are socially used. Of course, for some, the physical location implicitly effects how spaces will be used, but as demonstrated, there are other variables that direct or influence movement.

The space of Mepas can be categorised as a shared space or an unintentional space of peace. As put forward by participants, the mall is popular for residents in the city and through this, the space facilitates interactions with 'the other' (C Interview 2015; J Interview 2015). In this capacity, Mepas is a space that, though staged in a one function, is used alternatively as individuals can unintentionally socially rescript conflict divisions. This may be in a direct or indirect way, but what can be noted is the agentive capability of individuals to rescript space in subtle ways, which constitutes a conflict transformation in the city. For example, an individual would probably not go to the mall to interact across ethno-nationalistic divisions but, since there is only one such mall in the city, they may find themselves unintentionally doing so. In this capacity, functional movement to the mall can facilitate engagement across ethno-nationalistic divides. This provides the opportunity for what Mac Ginty (2014: 555) puts forward as 'everyday diplomacy'. This transforms not only social relations but spatial ones, as this creates new experiences that transform space and rescript of narratives of the ethno-nationalistic division. The aspect of unintentionality in social movement to the mall is of particular significance as the space can provide interaction across ethno-nationalistic divides, without the movement being weighted with the "pressure" of renegotiating conflict divisions. Fundamentally, such movement has the potential to not only rescript experience of space but also rescript preconceptions of other social actors, who also use that space.

Critically, the space of the mall facilitates transgenerational usage, from parents or guardians taking children to the play facilities in the mall, to the elderly using the space for heat in the winter, and air conditioning in the summer. In this latter capacity, the mall represents a functional space for older generations. However, the purpose behind the movement represents a transgenerational economic issue that can be read as crossing ethnic divisions. This is also identifiable with respect to the costliness of shopping at the mall, this sets the scene for the popularity of the humanitarian auctions which allow for women to get a new item of clothing for a reasonable price. While the collection of proceeds for someone who needs to fund a surgery further illustrates the untenable economic situation in the city. These varying modes of engagement mean that the mall is not only an open space, or a functional space, but one that is temporally rescripted through social usage and malleable to the needs of the intersecting social groups that use the space. This demonstrates two things about space but more importantly the interaction of social actors and space. Firstly, the purpose of the space can be transformed socially. Secondly, multiple overlapping narratives can be scripted in one location.

Locally Scripted and Staged Spaces

As we have seen through the above spaces, social actors have been able to transform city space through movement, or locally led interventions. A variety of spaces in the city foster shared space for residents to move through and also transform through movement. To summarise, the Mostar Rock school has been locally established and promotes inter-ethnic interaction of students through the formation of ethno-nationalistically diverse bands. The space of the Mostar Rock School is linked to the UWC, OKC Abrašević and the Street Arts Festival through cooperative activities and a shared ethos of promoting and facilitating inter-ethnic space. In many respects, these organisations embody a network of organisations in the city that transform the use of the city space for those who use these spaces. They do so through facilitating interaction across ethno-nationalistic divides and through shared spaces. These shared spaces create social networks which transcend ethno-nationalistic divisions and have fostered movement across the city. Notably, participants focused on narratives of locally scripted or locally staged places, which were of more significance to participants than internationally staged spaces. Furthermore, the location of spaces in the city, does not, for all, mean that the space is an exclusively Croat or Bosniak space. This was demonstrated through the Old Glass Bank and the Kosača cultural centre. However, some participants, J and B, criticised the organisation of the Kosača centre for prioritising political interests and only inviting certain performers, respectively. Similar criticisms were also raised regarding the prevalence of divisions in the youth services and clubs in the city, indicating the persistence of divisions. However, it is significant to note that, despite a persistence of ethno-nationalistic divisions there is a tangible personal enactment of conflict transformation fostered by participants. This was enacted through organised activities, or social movement which rescripted their experience of the divided city space. As a result, a peace network in the city, composed of OKC Abrašević, the UWC and the Rock School among others is observable due to the interactivity of these spaces. However, due to the social composition of spaces, those which are intended to be open, may in fact, for some be closed, this may be due to education, personal interests or subtle signifiers of social identity which script the space.

Additionally, the discussion of tourism that surrounded the Old Bridge presented divergent perspectives of the use of public space. Despite the international narratives surrounding the bridge, as a symbol of reconciliation participant's discussions categorised movement to the Old Bridge as typically functional. While some participants criticised the commercialisation of the space, the potential of the space as a source of income was unfulfilled for others. In particular, an entrance fee to the UNESCO heritage site of the Old Bridge was discussed by participants K and N, in order to raise funds for the city (Interview 2015). The economic issues raised concerning an entrance fee to the bridge can be linked to those raised by a majority of participants who highlighted the vulnerable socio-economic position of many in city and wider country.

While there are divergent narratives of the Old Bridge, there are other more visibly divisive spaces in the city, in particular, the Partisan memorial which was a popular space for numerous participants. The memorial has previously been in a state of disrepair, and a battleground for graffiti of fascist and anti-fascist groups. Ideologically, it had been a contested space; which C, D, J and K linked to institutional divisions. The spatial location of the memorial in a Croat city area was considered causal to the staging of the memorial as derelict. However, the existence of the Facebook online social network, 'Partisansko spomen-groblje – Help to preserve famous WW2 memorial in Mostar' exemplifies the social efforts to rescript the staging of the memorial, while the incorporation of the memorial into a City of Lights festival (by participants' H and G) demonstrates the agentive capability of social actors to directly rescript the use of space, even temporally. Fundamentally, the local efforts to preserve the memorial gained traction in 2018 through the inclusion of the memorial in the city budget and the April 2018 reconstruction activities which have increased the footfall to the memorial (City of Mostar Budget 2018; Šotrić 2018; Bljesak.info 2018). This demonstrates that social actors can raise awareness and promote reconstruction projects, in order to instigate the institutional funding and support which is generally necessary for large scale renovation, and certainly for the maintenance of contentious spaces.

In 2003, Norbert Winterstein (2003: 4) characterised the Kosača cultural centre as a 'blatant example of division' in the city of Mostar. Thirteen years following this Kosača remains a divisive space in the city. It can be observed through participants that the extent to which the space is considered divisional depends on the individual negotiating the space. Nevertheless, it is identifiable that the cultural centre, though used by the UWC as a location for graduation, exists as a space imbued with symbols that reflect the conflict divisions in the city (through the inclusion of the HVO flag alongside the Croatian flag and the crucifix sculpture in the courtyard). Notably, the use of the centre for UWC student graduations, rescripts the narrative of the space, however, it is significant that none of the younger respondents regardless of ethnicity used the space as a meeting point, and it was only appealing to one older and self-identified Croatian national. Critically, though not explicitly causative, that there is a lack of transgenerational appeal of the space which is arguably ethno-nationalistically charged. The type of musical performances on at the centre returns us to an earlier point about the generational appeal of certain types of music, and how this can connect certain individuals to certain spaces but also ideologies. As demonstrated, during the escalation period of the Bosnian War, Rock 'n' Roll became the 'moral' counterpoint to increasingly divisive political discourse and politically charged traditional 'folk' music (Kaneva 2011: 214). This is not to say that all folk music is politically divisive, or to read into non-movement to Kosača as a rejection of nationalistic discourse, but to note that there are different transgenerational responses to music and therefore cultural heritage. This is not limited to post-conflict spaces but in such spaces this could direct movement away from ethno-nationalistically divisive spaces which may only promote one narrative.

SPACES OF INTENTIONAL AND UNINTENTIONAL TRANSFORMATION

OKC Abrašević, can be considered both an intentional and unintentional space of transformation; as the centre requires no cost or educational subscription for movement, in this way it exists as a unique shared space in the city, the centre works beyond its spatial location, and actively challenges the ethno-nationalist cityscape. The centre engages with the city space in conceptual and physical ways, in particular, OKC Abrašević has facilitated initiatives such as the Street Arts Festival and has cooperated with other local organisations. In this capacity, the centre can be considered to be a key nexus of the shared spaces in the city. Though the space exists as an intentional shared space, it is worth noting that no space is entirely open to all, due to the social scripting of space which impacts on the perceived social rules and therefore use of space. In focus subtle codes of social membership or spatial affinities direct social engagement in the space. However, the centre has established, maintains, and supports a growing network of like-minded initiatives and organisations. What differentiates the space of Abrašević from other spaces in the city, is the unique concentration of social rescripting which characterises the establishment and maintenance of the centre. Through the establishment of Abrašević, the capability of individuals to socially rescript space is evident. As demonstrated through the narrative of the establishment of the centre, institutional and monetary capabilities of governmental and top down actors facilitate the restaging of space. However, the success and longevity of the centre underscores the capability for social actors to transform meaning in space. Therefore, the centre is an important site in consideration of the post-conflict involvement of social actors in conflict transformation and demonstrates that social actors can rescript institutionally staged space.

When setting out the concepts of staging and scripting in the first chapter, I highlighted the example of a shopping mall and a park, as spaces which are staged in one function but used in another function. In focus, narratives of movement in Mepas Mall in Mostar demonstrate the capability of social actors to rescript the meanings of staged space through movement. Due to the uniqueness of the space, it is widely used by residents and the mall can be considered a shared space in the city, in particular, the space functions as a positive space to see "the other". The normalisation of the unfamiliar other is important in the process of post-conflict transformation and represents the ability of social actors to rescript not only space but social interactions therein. Mepas Mall is observably open to all, though not set out as a space to build peace, the mall is an unintentional space of peace. It is unintentional as the mall is not staged for the purpose of traversing ethno-nationalistic divides in the city; but due to the widespread appeal of the space, it inevitably draws separated groups together, facilitating conflict transformation.

While the link between affordability, interaction in spaces, and rescripting experiences is evident, the humanitarian auctions, led by an ethno-nationalistically mixed group of women, rescripted the mode of consumption in the space of the mall. The auction group facilitated, not only socially generated financial support for those in need, but meant that the women involved could purchase an item of clothing they may want. Overall, while the space is staged as a mall and a site of consumer capitalism, the social use of the space indicates a nuanced utilitarian narrative of the space. This can be seen transgenerationally, as elderly residents use the space to save on heating and benefit from air conditioning. Furthermore, the unintentional function of Mepas as a shared space is important as the interaction with the other is casual and coincidental, in that it is not a concerted effort to "make peace"; but the capability to rescript and transform divisions through interactions with the spatial other is evident in this type of movement.

Holding Mepas Mall in comparison with OKC Abrašević, both spaces can be regarded as open to all, and the potential economic cost of engagement (which is potentially no cost as you could visit both spaces without spending any money[7]) with both spaces, can be regarded as similar. However, reflecting on the perceivable meaning of movement to the spaces, for some, movement to OKC Abrašević may mean engagement in a space bound up with ideological narratives that echo the divisions of two conflicts in the city, and may also represent an intentional movement to engage in a space of peacebuilding. However, social movement to the mall can be regarded as being unbound to the concept of peacebuilding or conflict transformation, though it holds the potential to facilitate interaction with "the other". Therefore, social movement to the mall unintentionally promotes a rescripting of social divisions of movement and use of space, which facilitates conflict transformation.

The narratives of participant movement and use of the city space demonstrates the spatial agency of social actors in rescripting staged space. The subtle and sometimes unseen rescripting of space, changes the use of space in ways that can cross ethno-nationalistic divides. It is through this socially led, though perhaps unintentional, process that conflict transformation is observable through social movement which rescripts use of the divided city space. However, as seen in the example of Abrašević, the support of institutional actors of restaging is important in the establishment of space. Though Abrašević demonstrates that such spaces can remain largely autonomous to the institutional actors, through prioritising the social scripting of the space.

NOTES

1. When I went to the museum in the middle of the day and at the weekend I was the only person under the bridge and in the tower. It cost roughly 5 km to enter but was less visible than the bridge itself and the views, though interesting, were less picturesque than the elegant arch of the Old Bridge.
2. In Cape Town, 'Kaapse Klopse', developed the use of humour and parody in its carnival-like parade which is held on the 2nd of January. Tweede Nuwer Jaar (Second new year) is rooted in the oppression of slavery and colonialism, after a brief period of being banned during the apartheid, the celebration remains popular in the city (South African History Online 2017).
3. Bogdanović had been the mayor of Belgrade in 1982 and was a critic of Serbian nationalism.
4. I term them neutral because they are not peace directed activities but by simply getting people in the same space, these activities perform a transformative function.
5. I make this distinction due to the international provenance of the brands and companies in the mall, in recognition as previously outlined, that all space is complex mix of global and local constellations of interaction.
6. As of 2016 the group no longer meets at the mall.
7. However, as previously noted Romani persons who sometimes beg in the mall are frequently escorted out by security.

BIBLIOGRAPHY

Amnesty International. (2017). *'We Need Support, Not Pity', Last Change for Justice for Bosnia's Wartime Rape Survivors*. Available from: https://www.amnesty.org/download/Documents/EUR6366792017ENGLISH.PDF. Accessed Feb 2018.

Benjamin, W. (1986/1928). Marseilles. In P. Demetz (Ed.), *Reflections: Essay, Aphorisms, Autobiographical Writings*. New York: Schocken.

Bljesak.info. (2018). *Vraća se stari sjaj Partizanskom, vratili se i turisti* [The Old Splendour Returns to Partisans, and the Tourists Return]. Available from: https://www.bljesak.info/kultura/flash/partizansko/234042. Accessed Apr 2018.

Bollens, S. A. (2007). *Cities, Nationalism and Democratisation*. London and New York: Routledge.

Björkdahl, A., & Selimovic, J. M. (2015). A Tale of Three Bridges: Agency and Agonism in Peace Building. *Third World Quarterly, 37*(2). Available at: https://www.tandfonline.com/doi/abs/10.1080/01436597.2015.1108825?-journalCode=ctwq20. Accessed 12 Feb 2016.

Bramsen, I., & Poder, P. (2018). *Emotional Dynamics in Conflict and Conflict Transformation.* Berghof Handbook for Conflict Transformation (Online Edition). Berlin: Berghof Foundation. First launch 15/02/2018. Available from: https://www.berghof-foundation.org/fileadmin/redaktion/Publications/Handbook/Articles/bramsen_poder_handbook.pdf. Accessed Mar 2018.

Calame, J., & Pašić, A. (2009). *Post-conflict Reconstruction in Mostar: Cart Before the Horse* (Divided Cities/Contested States Working Paper. No. 7). Available from: http://www.conflictincities.org/PDFs/WorkingPaper7_26.3.09.pdf. Accessed 20 June 2014.

Carabelli, Giulia. (2018). *The Divided City and the Grassroots: The (Un)making of Ethnic Divisions in Mostar.* Basingstoke: Palgrave Macmillan.

Cresswell, T. (1996). *In Place, Out of Place: Geography, Ideology and Transgression.* Minneapolis and London: University of Minnesota Press.

Croatian House of Herceg Stjepan Kosača. (2016). *About Us.* Available from: http://kosaca-mostar.com/web/onama.php. Accessed 15 Dec 2016.

Forde, S. (2016). The Bridge on the Neretva: Stari Most as a Stage of Memory in Post-conflict Mostar, Bosnia-Herzegovina. *Cooperation and Conflict, 51*(4). Available from: http://journals.sagepub.com/doi/pdf/10.1177/0010836716652430. Accessed 16 Dec 2016.

Gee, J. P. (2004). *Situated Language and Learning: A Critique of Traditional Schooling.* New York and London: Routledge.

Gee, J. P. (2005). Semiotic Social Spaces and Affinity Spaces: From the Age of Mythology to Today's Schools. In D. Barton & K. Tusting (Eds.), *Beyond Communities of Practice.* New York: Cambridge University Press.

Goffman, E. (1971). *The Presentation of Self in Everyday Life.* Harmondsworth: Penguin.

Hart, M. T. (2007). Humour and Social Protest: An Introduction. *International Review of Social History, 52*(S15). Available from: https://doi.org/10.1017/S0020859007003094. Accessed Feb 2018.

International Criminal Tribunal for the Former Yugoslavia. (2013). Prlić et al. judgement, Vol. 3, case no. IT-04-74-T, nos 764, 1581-3, 1585-6, pp. 459–461. Available at: http://www.icty.org/x/cases/prlic/tjug/en/130529-3.pdf. Accessed 5 Nov 2015.

Kaneva, N. (2011). *Branding Post-communist Nations: Marketising National Identities in the "New" Europe.* New York and Abingdon, Oxon: Routledge.

Kappler, S. (2013). Coping with Research: Local Tactics of Resistance Against (Mis-)Representation in Academia. *Peacebuilding, 1*(1), 125–140.

Kappler, S. (2014). *Local Agency and Peacebuilding: EU and International Engagement in Bosnia-Herzegovina, Cyprus and South Africa.* Re-thinking Peace and Conflict Studies. Basingstoke: Palgrave Macmillan.

Komora. (2016). *'Glass bank' in Mostar options for building the Federal Government?* Available from: http://www.komora.ba/vijesti/staklena-banka-u-mostaru-opcija-za-zgradu-vlade-fbih. Accessed 20 Mar 2016.

Kontra Press. (2013). *The Day When Barbarity Collapsed Old Bridge*. Available from: http://kontrapress.com/komentari.php?url=Dan-kad-je-varvarstvo-srusilo-Stari-most. Accessed 9 June 2016.

Mac Ginty, R. (2013). Conclusion. In R. Mac Ginty (Ed.), *Routledge Handbook of Peacebuilding*. Abingdon, Oxon: Routledge.

Mac Ginty, R. (2014). Everyday Peace: Bottom-Up and Local Agency in Conflict Affected Societies. *Security Dialogue, 45*(6). Available from: http://journals.sagepub.com/doi/full/10.1177/0967010614550899. Accessed 20 Oct 2016.

Massey, D. (1991, June). A Global Sense of Place. *Marxism Today*. Available from: http://banmarchive.org.uk/collections/mt/pdf/91_06_24.pdf. Accessed 20 June 2016.

Mostar Rock School. (2015). *History*. Available from: http://www.mostarrockschool.org/historijat.html. Accessed 5 June 2016.

Mostar Rock School. (2018). *History*. Available from: http://www.mostarrockschool.org/historijat_eng.html. Accessed Mar 2018.

Palmberger, M. (2016). *How Generations Remember: Conflicting Histories and Shared Memories in Post-war Bosnia Herzegovina.*, Global Diversities London: Palgrave Macmillan.

Participant A. (2015). Interview in Mostar.

Participant B. (2015). Interview in Mostar.

Participant C. (2015). Interview in Mostar.

Participant D and E. (2015). Interview in Mostar.

Participant F. (2015). Interview in Mostar.

Participant G. (2015). Interview in Mostar.

Participant G. (2016). Interview 2, follow up over email.

Participant H. (2015). Interview in Mostar.

Participant I. (2015). Workshop in Mostar.

Participant J. (2015). Interview 1 in Mostar.

Participant J. (2015). Workshop in Mostar.

Participant J. (2015). Interview 2, follow up over email.

Participant K. (2015). Interview in Mostar.

Participant K. (2016). Interview 2, follow up over email.

Participant L. (2015). Interview in Mostar.

Participant M. (2015). Interview in Mostar.

Participant N, O, P, Q. (2015). Group Interview in Mostar.

Participant R. (2015). Email correspondence.

Participant S. (2014). Informal Conversation.

Participant S. (2016). Email correspondence.

Partizansko spomen-groblje—Help to Preserve Famous WW2 Memorial in Mostar. (2015). *Facebook Page*. Available from: https://www.facebook.com/groups/147730451916207/. Accessed Dec 2015.

Pavarotti Music Centre. (2016). *About Us*. Available from: http://www.mcpa-varotti.com/index_eng.htm. Accessed 5 June 2016.

Ritzer, G. (2012). *The McDonaldisation of Society*. London: Sage.

Solis, G. D. (2016). *The Law of Armed Conflict: International Humanitarian Law in War*. New York: Cambridge University Press.

Šotrić, S. (2018, April 5). *Rekonstrukcija Partizanskog Spomen Groblja u Mostaru može otpočeti* [Reconstruction of the Partisan Memorial Cemetery in Mostar can Begin]. Available from: http://www.tacno.net/nasigradovi/rekonstruk-cija-partizanskog-spomen-groblja-u-mostaru-moze-otpoceti/. Accessed Apr 2018.

South African History Online. (2017). *The Cape Minstrels: Origins and Evolution of Tweede Nuwe Jaar [Second New Year] in the Cape*. First published, 11 December 2015, last updated 5 September 2017. Available from: https://www.sahistory.org.za/article/cape-minstrels-origins-and-evolution-tweede-nuwe-jaar-second-new-year-cape. Accessed Mar 2018.

Street Arts Festival Mostar. (2018, April 7). Facebook. *Čistka Staklene - buduće zgrade Vlade* [Glass Cleaning—The Future Government Building]. Available from: https://www.facebook.com/StreetArtsFestivalMostar/. Accessed Apr 2018.

UNESCO. (2005, July 15). *The Old Bridge Area of the City of Mostar*. World Heritage Scanned Nomination. Available from: http://whc.unesco.org/uploads/nominations/946rev.pdf. Accessed 14 June 2014.

Walasek, H. (2016). Cultural Heritage, the Search for Justice and Human Rights. In H. Walasek (Ed.), *Bosnia and the Destruction of Cultural Heritage*. Farnham: Ashgate.

Winterstein, N. (2003). *Commission for Reforming the City of Mostar. Recommendations of the Commission Report of the Chairman*. Available from: http://www.ohr.int/archive/report-mostar/pdf/Reforming%20Mostar-Report%20(EN).pdf. Accessed 20 June 2015.

Youth Council of the City of Mostar. (2018). *Initiative to Close the Entrance to the Glass Bank*. Available from: http://www.vmgm.ba/incijativa-za-zatvaran-je-ulaza-u-staklenu-banku/. Accessed Mar 2018.

Youth Power. (2018). *Snaga Mladih—YP—Youth Power*. Available at: https://www.facebook.com/pg/NGOYouthPower/about/?ref=page_internal. Accessed Feb 2018.

CHAPTER 7

Rescripting Spaces and Places: *Mostar and Other Divided Cities*

Understanding space, as fluid and infinite in its transformation, as a concept and as narrative experience holds unique potential for understanding conflict transformation processes. Social actors in divided cities are often written about as passive recipients of narratives of division. The ability to resist, or to transgress divisions is missing from many post-conflict studies of space. However, as demonstrated in the preceding chapters social actors can and do rescript narratives of space. It is important to note that social scripting can correlate with institutional divisions as discussed through the presence of divisional graffiti, and participant narratives of divisive movement. However, despite the power of institutional staging of space, the participant narratives of social movement in Mostar have demonstrated that the social scripting of spaces does not always correlate with institutionally staged space. Through social movement actors have rescripted narratives of divided space and social relations within space therefore enacting conflict transformation. Critically, conflict transformation may be time and context dependent due to the social actors using the space, though the impact of rescripting can be observed to be not limited to the spatiality, as it transforms not only relations of that space, but perceptions or expectations of other spaces. Critically, in so far as social actors are capable of transforming space, the institutional power of the staging and entrenchment of post-conflict divisions has also been demonstrated.

However, the two concepts are not strictly parallel in their spatial aims and there is cooperation between the two sometimes opposing concepts

© The Author(s) 2019
S. Forde, *Movement as Conflict Transformation*, Rethinking Peace
and Conflict Studies, https://doi.org/10.1007/978-3-319-92660-5_7

of rescripting and restaging. This has potential to be strengthened into a participatory approach to post-conflict reconstruction. Through participatory restaging, social actors can be involved discussions regarding investment, restoration, transformation, and use of social space in post-conflict spatialities. In Mostar, there are complex narratives of social movement through which social actors negotiate divided city space and relations with other social actors therein. This work has provided a discussion of the impact movement can have in entrenching or traversing divisions. As Massey notes (1991: 28);

> [i]f it is now recognised that people have multiple identities, then the same point can be made in relation to places. Moreover, such multiple identities can either be a source of richness or a source of conflict, or both.

In Mostar, the narratives of sub-divisions traversing the ethno-nationalistic divisions in the city demonstrates the 'richness' potential of diverse spaces. Furthermore, through sub-divisions of social movement, local actors have enacted conflict transformation through rescripting the ethno-nationalistically divided city space. While rescripting efforts, as performed by OKC Abrašević, the United World College (UWC), Mostar Rock School, and the Street Arts Festival are bound up in intentionality, the conflict transformation potential of unintentional transformation is also evident. This is the product of movement that is not for the sole purpose of fostering peace or rescripting narratives of space. However, such movement facilitates interaction with the other, and has the potential to rescript narratives of space and social relations within space. Through the social narratives of the use of the divided city space in Mostar, the agentive capability of social movement to foster conflict transformation, intentionally or unintentionally is evident.

DIVISION AND TRANSFORMATION

Though there are divisions of movement along ethno-nationalistic lines, for participants in this research, other divisions such as gender, ideology, class, and hobbies take precedence over ethnic divisions. Along theses constellations of social division, through social movement, residents of Mostar rescript meaning and usage of the divided city space. However, notably institutional actors and funding have an overriding influence of the staging of the space. The staging of space can be appropriated, as in the example of Mepas; or socially directed, as in the example of Abrašević. However, it can also translate to divisions over the usage of space, and as seen in the example of the Old Glass Bank and Partisan memorial. For these two spaces, the extent to which the institutionally supported, 2018 spatial transformation of these locations is maintained, will be revealed by time.

What can be taken away from this discussion of social movement in Mostar, is the capability of social actors to transform space in nuanced, subtle, and therefore, sometimes unseen ways. Additionally, it is through social movement in the divided city, that individuals interact with the 'other', such as when using the space of Mepas. Through such movement, individuals unintentionally engage in a shared space. Notably, movement to shared spaces, such as Abrašević may be dominated by the conceptualisation of the space as a shared space, with the aim of transforming society, which may deter some movement either from disinterest in the cause or pressure to do so.

Comparatively, the spaces of OKC Abrašević and Mepas demonstrate that regardless of staging, all space is social and has socially inscribed characteristics which may promote or deter the use of the space. Nevertheless, the functionality of Mepas means that while participants may not have fully positive feelings about the space, they still move through, and use the space. While the roster of activities in Abrašević is expansive and engaging; the cinema, grocery store, international brands and the air conditioning at Mepas hold a greater appeal for a wider audience, which therefore translates to a higher footfall of social movement (Ethnographic research, 2015). Notably, the impact of such movement may be weighted differently. For example, engagement with Abrašević led to the founding of the Street Arts Festival, while movement to Mepas has led to familiarisation with 'the other', observably the former may have a wider social impact than the latter (C Interview 2015; J Interview 2 2015). These interactions, which are hard to weight comparatively, could both offer an opportunity for a transformation of stagnant views of the other. This transformation may be on a small scale of seeing the other and becoming familiar with the presence of the other, or on a larger scale that performs physical rescripting of the city space. Both types of interaction demonstrate the agentive capabilities of social actors to rescript meaning in space. This critically demonstrates that social actors in divided cities have an opinion on the space and are not passive receivers of the divided city narrative established by political actors and jurisdictional divisions.

SPACES OF PEACE IN MOSTAR

It is of significance that while no participants felt the institutional divisions impacted on their personal movement, some participants' movement was effected by ethno-nationalistically divisive spaces in the city. Additionally, while the contemporary movement of participants was largely undeterred by divisions, for some, social movement had previously involved an element of division. In focus, prior to involvement in shared spaces, some participants had not moved to 'other side' of the city. Critically, while participants discussed their movement as occurring freely across the city, the overriding presence of the institutional division was evident. This is identifiable through an awareness of the divisions in education, politics, and also through reported discrimination due to the persistence of stagnated social divisions established by the conflict. As this research is spatially located, a large amount of data was produced regarding spaces of frequent movement, including cafes, shops, and other social spaces. Though multiple participants mentioned some spaces, there was no description of the movement, for example, movement to popular cafes, and clubs due to personal music preferences. While there was no descriptive narrative of such movement, beyond the categorisation that the spaces were enjoyable, it is notable that this movement typically took place on the 'Croat side' of the city. As previously outlined, the 'Croat side' of the city has a higher number of new cafes and shops. The concentration of such spaces is an important variable in directing social movement. Therefore, the categorisation of significant spaces of movement in the city of Mostar in this book was derived from the descriptive attention paid to the spaces. The spaces of Abrašević, the UWC and Mostar Rock School were frequently mentioned in terms of social movement. These spaces are also observably interactive in the city, conceptually and physically. Fundamentally, space and space identity are reliant on the social actors who use, and through use, script the space. In the establishment of Abrašević, local actors sought not only to offer a physical space which traverses the ethno-nationalistic institutional divides in the city, but one which offers an alternative space for engagement with the city space.

While presently, the centre acts as a conceptual hub in the city, in the future, Abrašević seeks to continue its work with education by offering an alternative platform for practical and social education (J Interview 1 2015). Though one of the main aims of Abrašević is to traverse ethno-nationalistic divides, the points of engagement which the centre provides also traverse sub-divisions, through their accessibility and wide appeal. The centre observably supports a network of like-minded institutions and supports the capability of civil society in the city of Mostar through the support of individual projects, which engage with the city space. This is most notable in two projects which stem from members of the OKC Abrašević centre, the Street Arts Festival and also the City of Lights project. The City of Lights Festival temporally rescripted the city space through providing physical illumination to spaces (such as the Partisan memorial) in the city which would not usually be accessible at night. While the Street Arts Festival annually provides a tangible way of engaging with other residents in the city through art and artistic workshops. In focus, the Street Arts Festival, established by Abrašević members, transforms spaces across the city. In particular, the Old Glass Bank, in its derelict state, became a blank canvas for artists involved in the festival to interactively display their work. Aside from physical transformation of the city, the festival engages local youth in the city space through creating a temporally dependent shared space across the city, therefore rescripting the divisive city space. As a result, the festival encourages shared use of public space, which, through social movement, facilitates conflict transformation. Fundamentally, the Street Arts Festival demonstrates how social actors can transform space through art which influences movement.

While the festival demonstrates rescripting which traverses the divisions in the city, graffiti around the city socially scripts the space as divisive. In focus, through the visibility of fascist and anti-fascist graffiti, an ideological sub-division is physicalised in the city. While discussions which mentioned the presence of an ideological division noted this did not impact upon movement, it can be regarded as relevant that the presence of the graffiti (Antifa and fascist) mirrors the division lines of the 1992–1995 conflict. The ideological graffiti rescripts the use of space for one group over another. The visibility of street art and graffiti in public space represents social narratives of movement within the space, scripting the space interactively. Critically, graffiti sets out a narrative of ownership and expected use of the space and is frequently challenged by alternative narratives of space as demonstrated by the fascist and anti-fascist scripting in the city space.

SUB-DIVISIONS IN MOSTAR

Through narratives of movement in the city it is observable that an ethno-nationalistic divisive narrative is maintained in the city. However, while such ethno-nationalistic divisions are present in the city, sub-divisions of movement also exist in Mostar. These sub-divisions of movement are influenced by variables of identity such as class, gender, age, and ideology. These exist within the networks of individuals who seek to traverse the ethno-nationalistic divide in the city. Notably, such variables influenced movement to some, but not all, of the spaces which facilitate a shared use of space. As outlined in Chapter 2, spaces become places through social use of the space, movement and experiences script the social performance of physical space. Through this, it can be presented that social usage, scripts the social expectations of the space, which directs movement and use. All space has nuanced scripting which directs the use of the space, and this may appeal to certain actors over others. How we process and understand space, and therefore, how we interact with it, is dependent on our own personal spatial trajectories. Our spatial trajectories are dependent on aspects of our positionalities, such as class, gender, and age (to name a few) and how these are overarchingly effected by structural, cultural, and social powers. These variables impacted on movement in the following ways; economically, participants noted issues with engagement in spaces in the city—though this did not always prevent engagement—as seen in the example of the Humanitarian Auction group meeting at Mepas. Also the interactivity of the variables of economic resources and gender are illustrated by the example of the Humanitarian Auction group which rescripted consumption in the space of Mepas. The ethnically diverse group of women organised the auctions and raised money for individuals in the city, sometimes for surgeries (J Interview 2 2015; F Interview 2015). The existence and purpose of the group, to fundraise, in some instances for a surgery, demonstrates the untenable economic position for many in the city. Further reflections on the space of Mepas, which include reasons for transgenerational movement (by elderly residents to benefit from the heating in the winter) generates greater detail of the economic issues in the city (C Interview 2015; J Interview 1 2015). In Mostar, economic issues remain an important, and fundamentally divisive factor in the post-conflict space. Spatially, the economic issues demonstrate the interconnectivity of top-down actors in directing social engagement with city space. Overall, despite such boundaries, the dominant variable of movement for participants can be observed as functionality, but also their feelings towards a space. However, it can be analysed that, a lack of movement to different areas in the city due to functional movement could reinforce spatial divisions in the city. As discussed by H movement in shared spaces or open spaces can also be limiting and divisive (Interview 2015). Fundamentally, the scripting of some spaces in Mostar is ethno-nationalistically divisional and therefore correlates with the institutional staging of the city. However, as demonstrated the social use of space in the city transforms narratives of the city space, rescripting space and enacting conflict transformation of social relations through social movement.

Social Movement as Conflict Transformation in Mostar

Physically and conceptually, the spaces of the Rock School, the UWC and OKC Abrašević operate as spaces which seek to deconstruct the ethno-nationalistic divisions in the city. Through the maintenance of these shared spaces, it is discernible that there is a persistence of ethno-nationalistic division in the city. The complexity of spatial narratives is also demonstrated by the spaces of Abrašević, the UWC, and Mostar Rock School, which function as shared spaces, as a close reading demonstrates that the spaces operate as both open and closed in different ways. Primarily, this is because the UWC, though active in the community, is an educational facility for youth. Additionally, Mostar Rock School is aimed towards youth (mid to late 20s), who want to learn to play rock music. Therefore, access and use of the space is dependent on having an interest in rock music, learning rock music, and being able to afford the classes. While Abrašević, can be observed as providing different 'spaces' of interaction within the one physical site. Those who worked to establish the centre have created not only a physical space of shared movement; but a hub which fosters contact with other local actors who interact and generate new shared spaces. The Centre is open to all (without age or education restrictions), however, like all spaces, Abrašević is a closed space for some in the city. This is due to the ideological origins of the centre and is irrespective of the varied programme of workshops, gigs, talks, film screenings, book launches, and other activities hosted there. Notably,

OKC Abrašević is an intentional shared space, one that is created and maintained socially for the purpose of overcoming ethno-nationalistic divisions in the city. Furthermore, as a socially formed space, though institutionally funded, OKC Abrašević also presents an important case for the value of participatory involvement in post-conflict spaces. In particular, the scripting of the Abrašević demonstrates the capability of social actors to operationalise spatial transformation and maintain shared spaces.

In contrast to OKC Abrašević, is Mepas Mall, as highlighted by participants, the space is used in a variety of ways which rescript the space of the mall. This was evidenced through the use of the mall by elderly residents for the benefit of air conditioning or heating (weather dependent); and the use of the space of the mall for the 'Humanitarian Auctions' organised by an ethnically diverse group of women. Alongside these examples of rescripting, the mall can be considered to be a shared space through the normalisation of interaction with 'the other'. This rescripts not only the space of the Mall, in a Croat city area, but also has the potential to rescript social relations within the space, by familiarising people with the spatial other. Through the shared nature of the space, social movement to the mall may not be intentional for the purpose of conflict transformation, however, intentionality does not impinge upon the rescripting of spatial and social experiences. Therefore, unintentional conflict transformation can occur through social movement.

The spaces of Abrašević and Mepas can be held comparatively, as both spaces are without any divisive national staging.[1] However, while both spaces offer up a physical location which is open to all citizens, the two spaces represent divergent narratives of the use of, not only social space, but narratives of engaging with others in social space. The space of Mepas and Abrašević are comparable regarding economic accessibility (the price of coffee at both locations is the same). However, Mepas is representative of consumer capitalism, whereas the space of Abrašević has roots in a socialist past. In this capacity, Abrašević and Mepas are contrasting spaces of spatial movement but also represent two contrastable ways of engagement with, and in, public space. The modes of engagement theoretically script ideological spaces over physical spaces. In particular, the mall represents capitalism and temporally exists as a present and potential future embodiment of space in the city. Conversely, the space of OKC Abrašević, due to the socialist roots of the space, represents the history and contemporary struggles of the city and overall, the centre demonstrates the agentive capability of social actors to transform post-conflict space.

MOSTAR AND OTHER DIVIDED CITIES

Some important lessons can be taken from Mostar and from this exploration of social movement in the city. Firstly, the concepts of rescripting and restaging, illustrate the importance of interdisciplinary approaches to post-conflict research. Through understanding the social construction of spaces they can be used as theoretical constructs to map the narratives of social movement in (divided) cities, as the framework lends itself to any context wherein individuals use space.

Secondly, the detailed discussion of social movement in Mostar has been facilitated through the prioritisation of social narratives as the basis for the exploration of the city space. In Mostar, there are multiple sub-divisions which transgress the dominant ethno-nationalistic division in the city. Furthermore, through a close reading of the mapped spaces and contextualisation through social narratives, the locations discussed can be unpacked from their East or West spatial position. This book demonstrates the need for future research into social movement in divided cities, to provide a better understanding of the use of space, including, but not exclusively focused on conflict divisions. The observation of the agentive capabilities of social actors is crucial in divided post-conflict space as it materialises the subtle ways in which conflict transformation can take place, intentionally or unintentionally.

Fundamentally, the applicability of socially led narratives and mapping of space through cartographies of transformation, is evident in other divided cities, for example, the city of Belfast is not only spatially dislocated, with a divided education system, but also has "peace walls" or boundaries which intersect communities. The 1998 Good Friday agreement brought an end to the conflict in Northern Ireland, with the conflict divisions roughly along the lines of 'Catholics v. Protestants, Republicans v. Loyalists, Nationalists v. Unionists' (Calame and Charlesworth 2011: 62). Peace walls were erected at interfaces to deter violence between communities during the conflict, with walls also built following the Good Friday Agreement (Prieto 2015). In a 2015 study, on attitudes in Northern Ireland towards the peace walls, the proportion of respondents[2] wanting the peace walls to remain increased 'from 22% in 2012 to 30% in 2015' (Bryne et al. 2015: 31). While there was a decrease noted in those who want the peace walls to 'come down some time in the future...from 44% in 2012 to 35% in 2015' (Bryne et al. 2015: 31). The results of the survey demonstrated a disconnect between institutional plans to restage the space and the social rescripting or perceptions of the walls. This is exemplified by the government plans to remove the peace walls, while social relations have 'deteriorated over the last three years' (Bryne et al. 2015: 31). Significantly, a lack of local agency in the post-conflict deconstruction of the peace walls is evident, with an increase in respondents from 28% in 2012 to 44% in 2015 believing the 'drive to remove walls [is coming] from outside the local community' (Bryne et al. 2015: 31). This exemplifies the importance of local social actors' involvement regarding the transformation of post-conflict space. Notably, the existence, and perceived necessity by some respondents, of the peace walls themselves, can be discussed as a spatialisation of the divisive narratives of the conflict. It is therefore, important to include narratives of social rescripting of space in the interface zones, to discuss how conflict narratives have been transformed (if at all, and in what regard). In particular, resident narratives of movement can provide an insight to divisional movement, while locally involved approaches can help identify spaces which could have the potential to be socially scripted as shared spaces, which may over time lead to the deconstruction of the walls. Fundamentally, local involvement in post-conflict spatial transformation is important in establishing social rescripting which can contribute to long-term conflict transformation in divided communities.

Space, Time, and Transformation

This work has demonstrated how social movement enacts conflict transformation in the divided city of Mostar. In the institutionally divided city, conflict transformation is an ongoing process through social movement. While it is evident that ethno-nationalistic divides do pervade the city. These divides do not always dictate movement in the city space. Furthermore, the social narratives of movement presented demonstrate the impact of locally supported shared spaces which invite movement from all. Such spaces enable social movement and social rescripting which has, for many, transformed the divided city space, and conflict narratives of the spatial division. Fundamentally, involvement in one or more of the open spaces have rescripted participant perceptions of the wider city space and increased their own scope of personal movement in the city. This demonstrates the potential of social movement to enact conflict transformation through the agentive capabilities of social actors to rescript staged space and social relations within space. Through the process of rescripting, which may be intentional or unintentional, social movement facilitates a spatailised conflict transformation.

Consequently, reflecting on the social narratives produced through the research, the importance of a deeper analysis of post-conflict use of divided space is evident. This book calls to widen the discussion on divided cities through presenting the social narratives of transformation in Mostar. Fundamentally, the research presents the agentive capabilities of social actors, through their movement, to rescript, and to transgress institutional spatialised divisions, thereby enacting conflict transformation. Critically, this work proposes that the intentionality or unintentionality, of such movement is not causal to the effectiveness of rescripting, this applies not only to space but social relations within space. Though this work must conclude, the narrative of the use of the city space in Mostar, Bosnia-Herzegovina continues. Accordingly, this work encourages academic inquiry of social rescripting as a modality of conflict transformation at an everyday level. This work has evidenced that while institutional divisions prevail in the city, impacting upon education, politics, and everyday lives; social actors have fostered spaces of peace and have rescripted narratives of division.

Revisiting the metaphor of the mountain discussed by Massey (2005: 139) when researching post-conflict spaces, it should be kept in mind that change may be small, it may be momentary, it will probably not be linear, and ultimately, we (as researchers or practitioners) may be moving too fast to see it and for it to seem to be tangible. We must therefore seek to reorientate our understanding of space and time and accept that it may be our lens which fails to capture the transformation and not the transformation moving too slow or being ineffective. In essence, we should look beyond our temporally limited scope of understanding of conflict transformation and accept that some relations need a great deal of time and space in order to transform but may be doing so in subtle and unseen ways. Social movement is capable of enacting intentional or unintentional conflict transformation through rescripting space which is staged as divided, in the city of Mostar and, undoubtedly, other divided cities.

NOTES

1. Aside from the cinema in the mall being ran by a Croatian company and the subtitles for films in Croatian.

2. The study conducted by a team at Ulster University was sent to '4000 households and was completed by 1021' (Bryne et al. 2015: 5). Furthermore, the study also noted that 'more Catholics than Protestants participated' 59% and 32% respectively (Bryne et al. 2015: 8). The study attributes this to research by Murtagh and Shirlow from 2006 'which shows interface areas in Belfast have been increasingly populated by individuals from a Nationalist background' (Bryne et al. 2015: 8).

BIBLIOGRAPHY

Bryne, J., Gormley-Heenan, C., Morrow, D., & Sturgeon, B. (2015). Public Attitudes to Peace Walls: Survey Results, Ulster University, Department of Justice Northern Ireland Government. Available from: http://socsci.ulster.ac.uk/pws.pdf. Accessed 5 Nov 2016.

Calame, J., & Charlesworth, E. (2011). *Divided Cities: Belfast, Beirut, Jerusalem.* Mostar and Nicosia: University of Pennsylvania Press.

Massey, D. (1991, June). A Global Sense of Place. *Marxism Today.* Available from: http://banmarchive.org.uk/collections/mt/pdf/91_06_24.pdf. Accessed 20 June 2016.

Massey, D. (2005). *For Space.* London: Sage.

Participant A. (2015). Interview in Mostar.

Participant B. (2015). Interview in Mostar.

Participant C. (2015). Interview in Mostar.

Participant D and E. (2015). Interview in Mostar.

Participant F. (2015). Interview in Mostar.

Participant G. (2015). Interview in Mostar.

Participant G. (2016). Interview 2 Follow up over e-mail.

Participant H. (2015). Interview in Mostar.

Participant I. (2015). Workshop in Mostar.

Participant J. (2015). Interview 1 in Mostar.

Participant J. (2015). Workshop in Mostar.

Participant J. (2015). Interview 2, Follow up over e-mail.

Participant K. (2015). Interview in Mostar.

Participant K. (2016). Interview 2, Follow up over e-mail.

Participant L. (2015). Interview in Mostar.

Participant M. (2015). Interview in Mostar.

Participant N, O, P, and Q. (2015). Group Interview in Mostar.

Participant R. (2015). E-mail Correspondence.

Participant S. (2014). Informal Conversation.

Participant S. (2016). E-mail Correspondence.

Prieto, I. Á. (2015). *Northern Ireland Foundation: Peace Walls.* Available from: https://northernireland.foundation/sharedfuture/research/peace-walls/. Accessed 20 Dec 2016.

INDEX

© The Editor(s) (if applicable) and The Author(s) 2019
S. Forde, *Movement as Conflict Transformation*, Rethinking Peace and Conflict Studies, https://doi.org/10.1007/978-3-319-92660-5

221